# Emerging Materials and Advanced Designs for Wearable Antennas

Vinod Kumar Singh
*SR Group of Institutions, India*

Vikas Dubey
*Bhilai Institute of Technology, India*

Anurag Saxena
*SR Group of Institutions, India*

Ratnesh Tiwari
*Bhilai Institute of Technology, India*

Himani Goyal Sharma
*Department of Electrical Engineering, Poornima College of Engineering, Jaipur, India*

A volume in the Advances in Wireless Technologies and Telecommunication (AWTT) Book Series

IGI Global
PUBLISHER of TIMELY KNOWLEDGE

Published in the United States of America by
IGI Global
Engineering Science Reference (an imprint of IGI Global)
701 E. Chocolate Avenue
Hershey PA, USA 17033
Tel: 717-533-8845
Fax: 717-533-8661
E-mail: cust@igi-global.com
Web site: http://www.igi-global.com



Library of Congress Cataloging-in-Publication Data

Names: Singh, Vinod Kumar, 1977- editor. | Saxena, Anurag, 1989- editor. | Dubey, Vikas, 1985- editor. | Tiwari, Ratnesh, 1982- editor. | Sharma, Himani, 1973- editor.
Title: Emerging materials and advanced designs for wearable antennas / Vinod Singh, Anurag Saxena, Vikas Dubey, Ratnesh Tiwari, and Himani Sharma, editors.
Description: Hershey, PA : Engineering Science Reference, [2021] | Includes bibliographical references and index. | Summary: "this book discusses the latest research into how wearable antennas are being applied successfully in various part of life such as health monitoring, physical training, navigation, RFID, medicine, and the military with the aim of improving the quality of every day life"-- Provided by publisher.
Identifiers: LCCN 2020047982 (print) | LCCN 2020047983 (ebook) | ISBN 9781799876113 (hardcover) | ISBN 9781799884255 (paperback) | ISBN 9781799876120 (ebook)
Subjects: LCSH: Wearable technology--Antennas.
Classification: LCC TK7871.6 .E48 2021 (print) | LCC TK7871.6 (ebook) | DDC 621.382/4--dc23
LC record available at https://lccn.loc.gov/2020047982
LC ebook record available at https://lccn.loc.gov/2020047983

This book is published in the IGI Global book series Advances in Wireless Technologies and Telecommunication (AWTT) (ISSN: 2327-3305; eISSN: 2327-3313)

British Cataloguing in Publication Data
A Cataloguing in Publication record for this book is available from the British Library.

All work contributed to this book is new, previously-unpublished material.
The views expressed in this book are those of the authors, but not necessarily of the publisher.

For electronic access to this publication, please contact: eresources@igi-global.com.

# Advances in Wireless Technologies and Telecommunication (AWTT) Book Series

ISSN:2327-3305
EISSN:2327-3313

Editor-in-Chief: Xiaoge Xu University of Nottingham Ningbo China, China

**MISSION**

The wireless computing industry is constantly evolving, redesigning the ways in which individuals share information. Wireless technology and telecommunication remain one of the most important technologies in business organizations. The utilization of these technologies has enhanced business efficiency by enabling dynamic resources in all aspects of society.

The **Advances in Wireless Technologies and Telecommunication Book Series** aims to provide researchers and academic communities with quality research on the concepts and developments in the wireless technology fields. Developers, engineers, students, research strategists, and IT managers will find this series useful to gain insight into next generation wireless technologies and telecommunication.

**COVERAGE**

- Wireless Sensor Networks
- Global Telecommunications
- Telecommunications
- Mobile Technology
- Digital Communication
- Cellular Networks
- Wireless Broadband
- Broadcasting
- Radio Communication
- Network Management

IGI Global is currently accepting manuscripts for publication within this series. To submit a proposal for a volume in this series, please contact our Acquisition Editors at Acquisitions@igi-global.com or visit: http://www.igi-global.com/publish/.

The Advances in Wireless Technologies and Telecommunication (AWTT) Book Series (ISSN 2327-3305) is published by IGI Global, 701 E. Chocolate Avenue, Hershey, PA 17033-1240, USA, www.igi-global.com. This series is composed of titles available for purchase individually; each title is edited to be contextually exclusive from any other title within the series. For pricing and ordering information please visit http://www.igi-global.com/book-series/advances-wireless-technologies-telecommunication/73684. Postmaster: Send all address changes to above address. 

## Titles in this Series

*For a list of additional titles in this series, please visit:*
*http://www.igi-global.com/book-series/advances-wireless-technologies-telecommunication/73684*

***Principles and Applications of Narrowband Internet of Things (NBIoT)***
Sudhir K. Routray (Bule Hora University, Ethiopia) and Sasmita Mohanty (University of Aveiro, Portugal)
Information Science Reference • © 2021 • 436pp • H/C (ISBN: 9781799847755) • US $195.00

***Design Methodologies and Tools for 5G Network Development and Application***
P. Suresh (Vel Tech Rangarajan Dr. Sagunthala R&D Institute of Science and Technology, India) G. Vairavel (Vel Tech Rangarajan Dr. Sagunthala R&D Institute of Science and Technology, India) and U. Saravanakumar (Muthayammal Engineering College, India)
Information Science Reference • © 2021 • 291pp • H/C (ISBN: 9781799846109) • US $195.00

***Harnessing the Potential of AI for Modeling 6G Networks***
Moses John Strydom (University of South Africa, South Africa)
Engineering Science Reference • © 2021 • 300pp • H/C (ISBN: 9781799849698) • US $215.00

***Evolution of Software-Defined Networking Foundations for IoT and 5G Mobile Networks***
Sunil Kumar (Amity University, Noida, India) Munesh Chandra Trivedi (National Institute of Technology, Agartala, India) Priya Ranjan (Amity University, Noida, India) and Akash Punhani (Amity University, Noida, India)
Information Science Reference • © 2021 • 235pp • H/C (ISBN: 9781799846857) • US $195.00

***Managing Resources for Futuristic Wireless Networks***
Mamata Rath (Birla School of Management, Birla Global University, India)
Information Science Reference • © 2021 • 338pp • H/C (ISBN: 9781522594932) • US $195.00

*For an entire list of titles in this series, please visit:*
*http://www.igi-global.com/book-series/advances-wireless-technologies-telecommunication/73684*

IGI Global
PUBLISHER of TIMELY KNOWLEDGE

701 East Chocolate Avenue, Hershey, PA 17033, USA
Tel: 717-533-8845 x100 • Fax: 717-533-8661
E-Mail: cust@igi-global.com • www.igi-global.com

# Table of Contents

# Detailed Table of Contents



Micro-electro mechanical systems (MEMS) technology has facilitated the need for innovative approaches in the design and development of miniaturized, effective, low-cost radio frequency (RF) microwave circuits and systems. This technology is expected to have significant role in today's 5G applications for the development of reconfigurable architectures. This chapter presents an overview of the evolution of MEMS-based subsystems and devices, especially switches and phased array antennas. This chapter also discusses the key issues in design and analysis of RF MEMS-based devices, particularly with primary emphasis on RF MEMS switches and antennas.



Printed antennas are designed for various applications in recent years. The latest development of antennas are wearable antennas; these wearable antennas are playing a major role in the energy harvesting mechanisms, patient monitoring systems, and military applications. In this chapter a reconfigurable wearable antenna is proposed. Reconfigurable wearable antennas are able to operate in both modes on body and off body. These antennas are reconfigured from one frequency to another frequency depending upon the requirement. Three types of reconfigurable antenna modes are designed and analyzed in the chapter. Frequency reconfiguration with off body mode and on body mode, frequency reconfiguration between inside receiver and outside interrogator, and frequency reconfiguration with different substrate conditions. The choice of frequency bands are WBAN and other wireless application bands. Different switches are used to control the frequencies.

In the ongoing years, the radar innovation, once utilized dominatingly in the military, has begun to rise in various regular citizen applications. One of the regions that this innovation showed up is car automation. The danger of crash is incremented as the quantity of vehicles out. Versatile voyage control, pre-crash, and cautioning are a portion of the utilizations of car radar which are as of now being used in numerous autos. The eventual fate of the car business guarantees to offer a completely self-ruling auto that can drive itself with no driver help. These vehicles will require great radar sensors that can give exact data about the encompassing of the vehicle. These sensors will likewise require a figuring stage that can guarantee constant handling of the received signals. It is discovered that digital signal processing comprises of 3D FFT and framework augmentation handling. As an elective preparing stage, a FPGA/ASIC stage can be utilized to actualize of the calculation to meet the ongoing imperatives of car applications.



In this chapter, a diamond cut shape antenna has been simulated under CST software using twisting material (i.e., leather). The dielectric constant of leather material is 1.655. For antenna to be twist, leather substrate is used. The return loss of this newly design diamond cut shaped antenna is -35 dB at 5.5467 GHz resonant frequency. This type of antenna can be useful in BAN (body area network) applications. In this chapter, simulated results like return loss graph, 3-D pattern, and polar plot are presented.



The anticipated antenna is studied and presented. The anticipated antenna is connected to probe feed when designed on IE3D software. The bandwidth, gain, VSWR, and reflection coefficient have been found which is of 77.71% bandwidth, and gain is of 4.1 dBi. This antenna is compact, lightweight, and suitable for portable devices. By using some advanced techniques, the antenna provides dual bandwidth and triple bandwidth. The anticipated antenna provided large bandwidth for and is utilized for broadband applications.



In this chapter, artificial neural network is used for the estimation of bandwidth of a dual feed microstrip antenna. The MLPFFBP-ANN and RBF-ANN are used to implement the neural network model. The simulated values for training and testing the neural network are obtained by simulating the antenna on IE3D software. The results obtained by using ANNs and IE3D simulation are compared and are found quite acceptable, and also it is concluded that RBF network is more accurate and fast as compared to back propagation algorithm of MLPFFBP. The anticipated is applicable to operate in triple band from 2.208GHz-5.35GHz, 2.358GHz-2.736GHz, and 3.815GHz-5.143GHz. The antenna is also fabricated with FR-4 glass epoxy material. The experimental results, simulated results of IE3D, and simulated results of neural network are compared.



This chapter presents the various designing methods and challenges associated with wearable antennas, selecting the designing material, various fabrication techniques, and implementation methods on the fabric. Wearable antennas have gained popularity in recent years due to their unmatched properties and unique features. The wearable antenna is capable of providing effective wireless communication, tracking, and sensing. These types of antenna need to be conformal when used on numerous parts of the human body, and they need to be lightweight, flexible, and must be implemented easily on the flexible material. These antennas must be capable of operating with maximum efficiency on the human body. These requirements make the wearable antenna design challenging. They also need to be compact in size,

bi-directional power flow, light in weight, compact size, etc. They also compensate the environmental issues which are created due to transformer oil. Because of various advantages over traditional transformer, SST is preferred widely at the present time. So in this chapter, the various architectures, needs, and applications of solid state transformers are discussed. The global market of SST has continuously improved because it has several applications and benefits.



In the 21st century, energy supply has been one of the challenging issues that the world is facing. This is due to growing populations which results in more homes, businesses, and industries and has resulted in a large number of new appliances that increased the energy demand to a new level. The electric system we are using today is a one-way energy flow. Smart grid has emerged as the solution of continuously increasing demand. It is the future of electric energy system and also the power grid for upcoming generations. SG system works on the dual technique, that is, it counts electricity flow and the information which could deliver and distribute the energy universally by automating the existing electrical network.



In the Chhattisgarh state of India, the Bastar region is famous for its various forest produce collected by local inhabitants. These collected forest produce are dried in open sun drying. This chapter presents the design and analysis of solar dryer which is simple in construction and a low-cost dryer. The construction of the solar dryer will be made of galvanized mild steel sheet absorbing the sun's radiation. The hot air will be taken out through a wind turbine ventilator at the top to maintain the required humidity level in the solar dryer. The analysis shows that the temperature inside the solar dryer is evenly distributed. The suggested solar dryer will reduce the drying time of the forest produce.


*Aloke Verma, Kalinga University, India*
*A. K. Diwakar, Kalinga University, India*
*R. P. Patel, Guru Ghasidas University, India*

In this chapter, CH3CH2NH3SnI3/TiO2 heterojunction lead-free perovskite solar cell (HLFPSCs) is produced by the deposition of perovskite nanoparticles from a solution of CH3CH2NH3I and SnI3 in γ-hydroxyl butyric acid lactone on a 400 nm nanolayers of TiO2 showing (001) planes. The CH3CH2NH3SnI3 /TiO2 HLFSCs shows remarkable PV performance, with Jsc= 17.05 mA/cm2, Voc = 0.71 V, and FF = 0.61; resultant PCE of 6.5% in standard AM 1.5 G one-sun light intensity and PCE of 7.5% in lower light intensity was reported in this work. The beginning of such a simple solution treated HLFPSCs paves the way to appreciate low-cost, high-efficiency LFPSCs.


*Nikhil Agrawal, SR Group of Institutions, India*
*Dev Prakash Chourasia, Rajiv Gandhi Proudyogiki Vishwavidyalaya, India*
*Bharat Bhushan Khare, Rajiv Gandhi Proudyogiki Vishwavidyalaya, India*

This chapter presents a new single-phase multilevel inverter topology. This topology requires fewer components in comparison to conventional multilevel inverter topology such as diode clamped, flying capacitor, and cascade h-bridge inverter. In this topology, no extra clamping diode and no extra clamping capacitor are required for operation. This chapter also presents the control and modulation techniques. These techniques applied on this proposed MLI topology and simulation result are calculated with the help of MATLAB/SIMULINK software and are compared on the basis of THD with over and under modulation index.

# Preface

Wearable antennas have received plenty interest in current years because of their appealing functions and opportunities in allowing lightweight, flexible, low fee and transportable. Wearable antenna apparatuses are utilized for various applications such as Wide Band Area Network (WBAN), health monitoring. In particular, such necessities are needed to be considered on a need premise in the wearable reception apparatuses, for example, primary disfigurement, exactness, and precision in manufacture strategies and their size. Different specialists are dynamic in this field and, appropriately, some huge advancement has been accomplished recently. This book endeavors to basically audit the wearable antenna particularly considering new materials and creation strategies, and novel plans, for example, scaled down catch radio wires and scaled down single and multi-band reception apparatuses, and their special shrewd applications in WBAN. At last, the end has been attracted regarding some future bearings.

In Chapter 1, today's 5G communication systems, miniaturization has become a significant factor in the development of microwave integrated circuit components. With the advancement in MEMS/NEMS technology, micro/nano structures are being used. Various structural changes have been studied to achieve miniaturization. An elevated coplanar waveguide and space filling curves can be used to model a RF MEMS switch and comparing to conventional CPW structure confirming the possibility of miniaturization.

In Chapter 2, the inverted and implanted antennas are analyzed and presented here for wireless body area networks and other applications. This antenna can be radiated and reconfigured for minimum two frequency modes in every switching condition. At every switching condition one frequency mode is operating for WBAN and the second mode is operating for other wireless frequency applications. Antenna parameters are simulated and analyzed for ON body and OFF body conditions with different position of the human body tissue. The antenna radiation parameters are analyzed for different substrate condition with high gain .A person, who can wear this antenna for receiving or transmitting the signal from body area networks and other wireless applications. This antenna also can be worn easily with cloths.

In Chapter 3, as the number of sensors increases the complexity of the signal processing increases. It is discovered that signal processing comprises of 3D FFT and matrix multiplication. Considering the calculation and the continuous necessity of the application, it was discovered that multi-processor system isn't skilled to meet ongoing imperatives of the application. As an elective handling stage, a FPGA/ASIC stage can be utilized to actualize of the calculation to meet the ongoing imperatives of automotive applications.

In Chapter 4, for BAN Network, a twisting material antenna has been designed and synthesis on a leather substrate with relative permittivity of 1.655. The suggested antenna has a return loss of -35 dB. The overall performance of presented antenna is better than the performance of square shape type antenna with FR-4 material. Further this antenna is used for BAN applications and can also be used for energy harvesting.

In Chapter 5, the proposed antenna design gives the bandwidth of 77.71%, gain of 4.1 dBi, and directivity is 5.2 dBi. The proposed antenna is simple, compact size and low cost and is suitable for broad band application.

In Chapter 6, a design of Dual Inset feed slotted cross patch microstrip antenna with line probe dual feed for WLAN and Wi-Max application is presented. FR-4 Glass epoxy substrate with dielectric constant 4.4 is used which gives a wide bandwidth of 83.10%, 28.12% and 14.80%. In this work ANN is used as a tool to study the frequency band and bandwidth of microstrip antenna. The results obtained from IE3D and those obtained from ANN are in good agreement and shows almost 99.6% with MLPFFBP and 99.99% accuracy with RBF. The training and test set has been designed with the data obtained from IE3D simulator.

In Chapter 7, the antenna manufacturers are working on the field of wearable technology since the last decade. We have seen the rapid growth in the last couple of years. Still, the expansion of wearable technology is quite slow in comparison with the other popular technologies. This is mainly due to less technological enhancement and also inadequate market needs. Recent popular technologies are motivated by the easy connectivity and need for data. A large number of people are comfortable to use technology and can spent on the things that enhances their lifestyle and quality of life. Easy communication and tracking devices and sensors are being famous day by day.

In Chapter 8, the proposed flexile antenna has been designed and simulated successfully for enhancing bandwidth. The anticipated antenna has been simulated in the frequency range of 2 to 8 GHz having bandwidth of 106.30%, and directivity is 3.609dBi at 3.2 GHz & 4.519 dBi at 4.8 GHz. The proposed antenna is simple, compact size and low cost as less copper will be needed to fabricate the patch and ground. Due to satisfactory performance of antenna it has many applications like in telecommunication, aircraft surveillance and different wireless communication systems.

In Chapter 9, a novel design of bendable flexible textile antenna has been designed using textile material like jeans. The CST microwave studio is used to model the anticipated antenna and after simulation various results are obtained. The anticipated antenna has been provided the bandwidth from 1.90 GHz to 6.81 GHz having wide bandwidth of 112.74%. The designed antenna is suitable for wideband and Bluetooth application.

In Chapter 10, Solid State Transformer plays an important role in all three basic components of power system i.e. generation, transmission & distribution. It can overcome the constraints of conventional transformer from the point of view of weight, size, harmonic suppression, current limiting. Researchers have proposed different topologies and architecture of SST to improve the efficiency and reducing the size moreover they have designed smart transformer by connecting real time monitoring system in SST.

In Chapter 11, we can see with the growing years it is the need to maintain our systems smartly. They will support the existing system and will improve the power system networks. As we have seen in this paper Smart Grid systems are really a boon for the existing systems. As they will maintain a two way process between the utility end to the consumer end. SG system will reduce the losses of the line as a result efficiency will be improved, distribute the power, reduce the emission, have a good control on the consumption, proper communication at each end and whole electric system. It will also make the consumers aware of the electric energy systems. So, all nations of the world are working for making their electric systems smart so that electric energy supply should be done properly and reliably amidst of increasing demand.

In Chapter 12, as the collected forest produce definitely becomes valuable and useful if dried and stored for a long time. This ensures the even distributions of temperature and hot air velocity inside the solar dryer. The selected and tested designed dimension also can be easily moved. This solar dryer gives local inhabitants to improve their GMP (Good Manufacturing Practices). That can be manufactured an enhanced to stored valuable dried collected forest produced to reach at a larger population in future.

In Chapter 13, the work successfully reported a nano-crystalline based on Hetero-junction Lead-Free Solar Cell (HLFPSCs) and crystal structure of the synthesized $CH_3CH2NH_3SnI_3$ was determined. Under full sun illumination, the $CH_3CH2NH_3SnI_3$/$TiO_2$ hetero-junction lead-free solar cell showed an overall conversion efficiency of 2.8%.

In Chapter 14, a new MLI topology presented for seven level in this paper. In this paper a comparisons are made between the proposed topology and conventional topologies in symmetric manner. This comparison study shows that proposed MLI topology require less number of switches count. The performance of proposed topology is analyzed by MATLAB/SIMULINK 2010b software. In proposed topology the THD value is obtained is nearly approx. to IEEE standard.

In Chapter 15, the presented proposed badminton shaped antenna has designed by using a textile material Jeans with a thickness of 1mm.The antenna operates at triple band frequency configuration in the C-band and X-band. The frequency range of the designed antenna is in (6.8GHz-7.6GHz) for C- band and (7.9GHz - 8.05GHz and 8.1GHz -10.35GHz) for X-band applications. The directivity of the proposed antenna is 4.574dBi, 4.593 dBi and 4.488 dBi at resonant frequencies 7.335 GHz, 8GHz and 8.38GHz respectively.

As this book covers relatively wide areas and numerous contents connected to complex scientific issues, errors and omissions may be unavoidable due to the limited knowledge and competence of the authors, therefore we sincerely appreciate the criticism and comments from the readers. We feel confident it will be highly valued for its documentation of the important field of optimization of sensors and antennas for wearable device and its practical applications. The editors wish to thanks all the authors who have contributed to this book and sincerely thanks to the IGI-Global Editorial and production team for the constant support. We send our best wishes to our readers!

*Vinod Kumar Singh*
*SR Group of Institutions, India*

*Vikas Dubey*
*Bhilai Institute of Technology, India*

*Anurag Saxena*
*SR Group of Institutions, India*

*Ratnesh Tiwari*
*Bhilai Institute of Technology, India*

*Himani Goyal Sharma*
*Department of Electrical Engineering, Poornima College of Engineering, Jaipur, India*

# Chapter 1
# Scope of RF MEMS Technology for Microwave Circuits and Systems

**Kanthamani Sundharajan**
https://orcid.org/0000-0002-6738-6353
*Tata Consulting Engineers, India*

## ABSTRACT

*Micro-electro mechanical systems (MEMS) technology has facilitated the need for innovative approaches in the design and development of miniaturized, effective, low-cost radio frequency (RF) microwave circuits and systems. This technology is expected to have significant role in today's 5G applications for the development of reconfigurable architectures. This chapter presents an overview of the evolution of MEMS-based subsystems and devices, especially switches and phased array antennas. This chapter also discusses the key issues in design and analysis of RF MEMS-based devices, particularly with primary emphasis on RF MEMS switches and antennas.*

## 1. INTRODUCTION

WITH the continual rise in today's 5G communications, there is an urge in the need for the development of a low-cost, multifunctional, adaptive, high data rate and low-power systems. Such systems are to be developed in such a way that prevent the deleterious outcomes of self intrusion due to multipath propagation and intervention from many people who intend to use the common channel. Additionally, it becomes essential to secure the information passed on from the people who have not planned and to reduce the outcome of jammers for the current and upcoming 5G wireless

DOI: 10.4018/978-1-7998-7611-3.ch001

systems. This crucial requirements forces to formulate a technology which leads to modified design and techniques.

Since RF front-end is supposed to act as a connecting media between a microwave system and its surroundings, it should comprise of highly efficient RF devices(Tsoulos, 2000) to find in a way around to meet the challenges of adaptability, effective use of spectrum, security, size reduction and cost minimization. Challenges in technologies including new architectures, reconfigurable, usage of newer materials and innovation in packaging techniques are essential to establish such RF systems (Peroulis, 2003). Hence there arises a need to develop such RF front systems from microwave to sub-millimeter-wave frequencies, since today's 5G communication networks are specifically designed for use in these bands.

The present semiconductor technology allows the development of RF transceiver components on a single substrate, which can be built using. diodes and transistors. Even though semiconductor technology is continuing to serve as the driving force for integrated circuit design and execution.(Nguyen et al., 1998),but today's RF systems require improved techniques to handle all sorts of data simultaneously(Lim et al., 2002). Enhancement in the existing systems or to replace, systematic investigation on Micro-Electro-Mechanical Systems (MEMS) is carried out everywhere in the world. MEMS have been recognized as the most enabling technologies in developing miniaturized, low cost and low energy RF components and systems.

This chapter gives an overview of the MEMS based subsystems and devices, taking into account different issues in respect of analysis and design of RF MEMS devices.

## 2. RADIO FREQUENCY MICRO ELECTRO MECHANICAL SYSTEMS (RF MEMS)

Advancement in MEMS technology had greatly influenced the present RF wireless communication systems with certain appreciable properties(Brown, 1999; De Los Santos, 2002; Katehi, 1995; Katehi et al., 2002; Petersen, 1979; Varadan, Vinoy, Jose et al, 2003) to move to GHz frequencies. Due to their specific advantages MEMS devices are preferred to transfigure the transmitter and receiver architectures.

With reduced power consumption attained via MEMS, design of RF passives and active circuits can be now achievable using high resistivity Si substrate, thus becomes cost effective. Micro-machining and MEMS technology promise to provide a novel approach in the evolution of effective and low-cost circuits and systems. They are expected to have significant application in developing low-cost and reconfigurable multiband antenna arrays. RF MEMS is an emerging sub-area of MEMS technology that promises to transfigure radio frequency and microwave applications. (Rebeiz, 2004).

Figure 1 (Nguyen et al., 1998) shows a typical wireless transceiver, making use of micro machining technique to replace the conventional blocks. Figure 1, indicates the possible replacement of off chip, high-Q passive components by micro machined versions.

Transmitting and receiving antenna size could be reduced using micromachining techniques and can be made reconfigurable for high frequency applications. The following sub-sections now gives an overview of possible existence of micro electro mechanical components that could be achieved using micromachining to improve the overall performance of the system.

*Figure 1. Typical wireless transceiver showing the possible replacement of MEMS components*
*(Nguyen et al., 1998)*

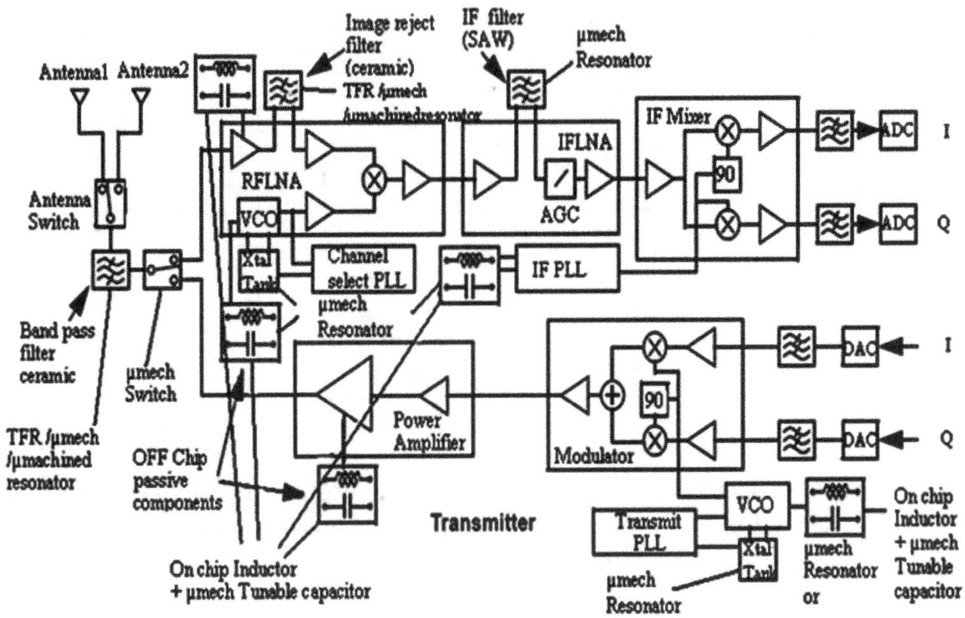


## 3. RF MEMS SWITCHES

RF MEMS exhibits as an emerging technology for certain GHz applications by reducing cost and improving performance(Lucyszyn, 2004). The performance of Radio-Frequency (RF) system for wireless communication application can be remarkably enhanced by the RF switches they use. RF switch is the basic and the most required component in these communication systems. RF MEMS switches are components that use mechanical movement to achieve a short circuit or an open circuit on the Radio frequency transmission line for switching the RF signal. Proven that RF MEMS switches have showed practically improved RF performance and figure-of-merit over the conventional PIN diode and FET based switches (Lucyszyn, 2004; Pranonsatit et al., 2006; Rebeiz & Muldavin, 2001).

Table 1 compares the solid-state switches built using PIN diodes, FET and MEMS switches. MEMS switches has higher level attributes compared to the existing solid-state switches in the respect to parameters such as insertion loss, isolation, and linearity, but has high operating voltage.

*Table 1. Comparison between solid state switches (PIN diode & FET), Electromagnetic Relay and RF MEMS switches*

| | **EMR (Hyman et al., 1999)** | **REED (Yao & Chang, 1995)** | **PIN Diode (Hyman & Mehregany, 1998)** | **FET (Goldsmith et al., 1998)** | **MEMS (Sovero et al., 1999)** |
|---|---|---|---|---|---|
| Insertion loss (dB) | 0.3 | 0.25 | 0.6 | 0.5 | 0.15 |
| Isolation (dB) | 25 | 15 | 25 | 15 | 35 |
| Switching-time (μs) | 1ms | 250μs | <1μs | 50ns | 70μs |
| Switching voltage /current | 5V/ 5mA | 5V/ 3mA | 5V/ 10mA | 5V/ 1μA | 68V/ 0.2μA |
| Contact resistance (Ω) | 0.15 | 0.13 | <1 | <1 | 0.15 |
| Return loss (dB) | >15 | >15 | 20 | >15 | 20 |
| Linearity (dBm) | >50 | >50 | <40 | <10 | >65 |

RF MEMS switches can be categorized into two groups, the shunt type and the series type, depending on the operating principle, as shown in Figure 2. A shunt-type switch normally remains in the 'ON' state and changes to the 'OFF' state when a control signal is applied. On the contrary, a series-type switch normally remains in the 'OFF' state, and changes to 'ON' state when a control signal is applied.

*Figure 2. Types of switches a) series connection b) shunt connection*

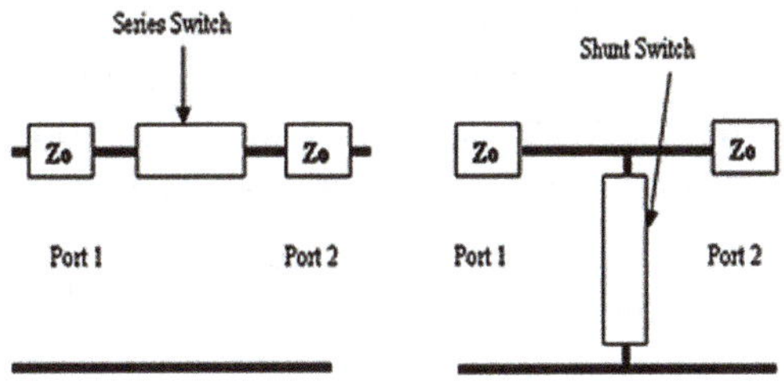


Depending on the method to transmit signal RF MEMS switches can be further divided as capacitive and resistive types as shown in Figure 3. The resistive type switches are realized by metal-to-metal direct contact, but in case of the capacitive type switches a dielectric layer is formed between two metals. This causes an impedance change between ON/OFF states.

*Figure 3. Actuation mechanism of a switch indicating ON-OFF states for a) Resistive b) capacitive switches*

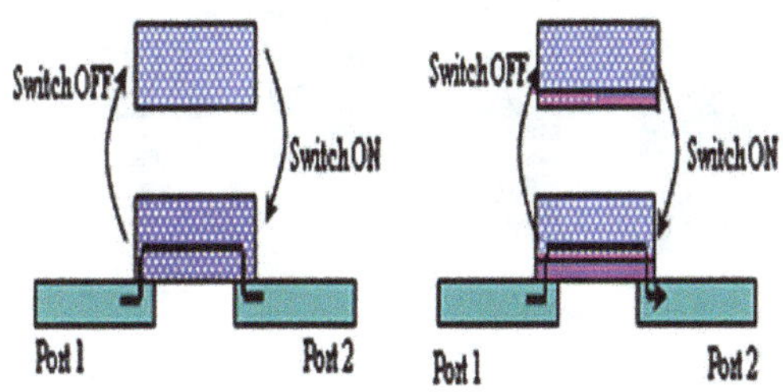


Based on the actuation mechanisms, the switches are classified as electrostatic, magneto static, piezoelectric, and thermo mechanical types. Compared to other mechanism, electrostatic switches are preferred because of no power consumption, are fast but have a high-operating voltage .

Electrostatically actuated membrane-based switch built on silicon was first reported by Peterson (Petersen, 1979).MEMS switches subject to various actuation mechanisms such as electro-magnetic (Hosaka et al., 1994; Taylor et al., 1997; Tilmans et al., 1999),magneto static(Wright & Tai, 1999), electrostatic (Gretillat, Gretillat, & De Rooij, 1999), thermostatic (Sanders, 1998) and various structural designs including a rotating transmission line (Larson et al., 1991), surface micro machined cantilever beams (Schlaak, 1997; Suzuki et al., 1999; Yao & Chang, 1995), multiple membrane supported designs (Goldsmith et al., 1996; Muldavin & Rebeiz, 1999; Suzuki et al., 1999; Yao & Chang, 1995; Yao et al., 1999), bulk micro machined or wafer bonded designs (Komura et al., 1999), diamond material based cantilever (Admschik et al., 1999), poly silicon based switch(Gretillat, Gretillat, & De Rooij, 1999), mercury micro-drop membranes (Simon et al., 1996), and bi stable micro relays (Kruglick & Pister, 1998; Sun et al., 1998). Laterally actuated switches (Kruglick & Pister, 1999; Schiele & Hillerich, 1999) are also reported. Detailed review report of theses switches are presented in (Rebeiz, 2004; Varadan, Vinoy, & Jose, 2003).It is justified that MEMS-based RF switches could play an important role in 5G mobile communications (Iannacci, 2017; Iannacci et al., 2016a; Iannacci et al., 2016b). RF MEMS switches ranging from 10 MHz to 60 GHz with a reduction in pull down voltage are commending choice for mm-wave 5G applications(Shekhar et al., 2018)

*Figure 4. Switched filter banks for wireless applications to switch between Tx & Rx*

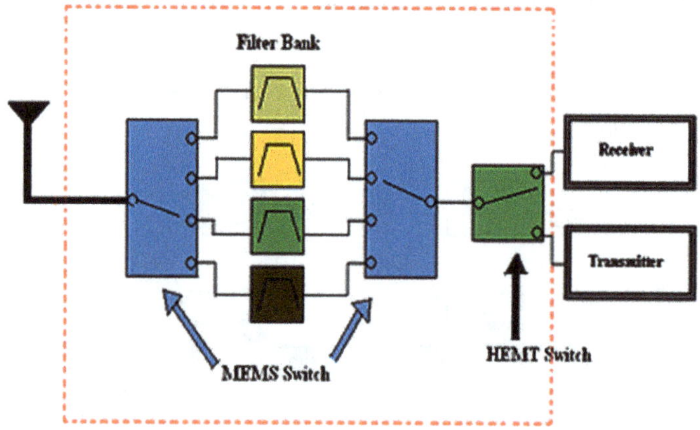


Various application areas of these RF MEMS switches are in tunable filter, adjustable amplifiers, variable Oscillator, and multi -band antenna. All these are usable for frequency band/mode selection to either transmit the signal to an antenna or to receive the signal from an antenna **(Cetiner et al., 2003)** as shown in Figure 4.

## 4. RF MEMS BASED PHASED ARRAY ANTENNAS

The 5G wireless communication systems require modern phased array antennas to meet out the objective of increased capacity, improved data rate and better quality of service. The phased array antenna is used to improve the performance for serving multiple users simultaneously by generating concurrent directional beams (Hong et al., 2017). Array of antennas, power dividers and phase shifters circuits constitute to form a phased array antenna as presented in Figure 5.

*Figure 5. Illustration of a phased array antenna showing the presence of power dividing network, phase shifters and antenna arrays*

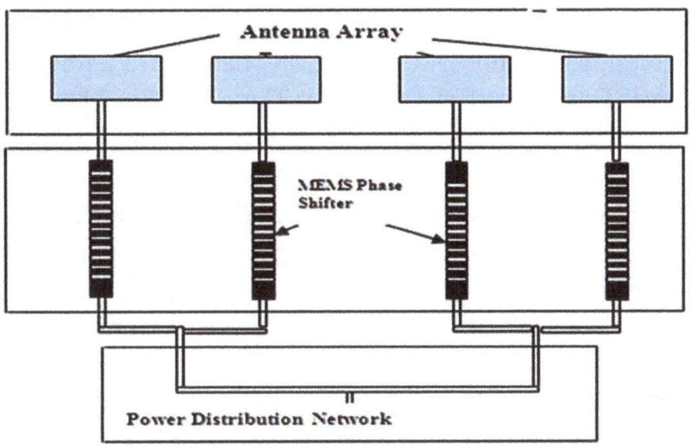


The phase shifting element is an important constituent of phased arrays for automatic beam steering .The beam steering can be mechanical or electrical. In mechanical beam steering the position of the antenna is varied using motors. Even though the mechanical beam steering is simple it is slower and creates noise in the system. The electrical beam steering is most preferred where faster beam tracking is required. In electrical beam steering phase shifter is used to vary the beam position rapidly (Real et al., 2018).

For the past decades the phase shifter designed using GaAs FET is very attractive because of its enchanced speed, very less power consumption and narrow band frequency (McLevige & Sokolov, 1980) but hard to maintain isolation and loss which depend on circuit configuration (Gaspari & Yee, 1978). The phase shifter designed using PIN diode consumes more DC power than FET but provide less loss performance. Due to light weight, smaller size, ferrite phase shifters are preferred in Radars but provide twist in cost and circuit complexity. The ferrite phase shifters are realized in waveguides, coaxial line, strip and microstrip line (Yao & Chang, 1995). The phase shifters are designed using a) parallel coupled transmission line of equal length to produce differential phase shift (Schiffman, 1958) b) switching between low and high pass filters producing broadband phase shifters (Garver, 1972) c) 3dB coupler loaded with two P-i-n diodes (Glance, 1979) d) Hybrid coupler (White, 1974) using varactor diodes by varying the junction capacitance (Kim et al.,

1966). These conventional phase shifters are lossy, consume more power and cannot handle more power at microwave frequencies (Miao et al., 2006). The integration of antenna element, conventional phase shifter and power divider in a phased array antenna is costly and introduce more losses. The RF MEMS technology is efficient in providing miniaturization and better solution in terms of performance(Guclu et al., 2010).

Due to performance enhancements such as reduced loss, enhanced isolation, low dc power consumption, high linearity RF MEMS phase shifters are preferred in wide band applications (Hayden & Rebeiz, 2002). The RF MEMS Distributed MEMS Transmission Line (DMTL) phase shifters using MEMS bridges which offers good performance was first proposed by(Barker & Rebeiz, 1998)

The capacitance of the loaded lines in the DMTL phase shifter can be increased by loading with MIM capacitor and MAM capacitor (Kim et al., 2002; Rebeiz et al., 2002). To use in modern communication system applications, phase shifters ranging from Ku to V band was designed using two back to back Single Pole Four Throw switches (Dey et al., 2017), Single Pole Eight Throw switches and Single Pole Sixteen Throw switches (Dey et al., 2018) to provide low loss, less phase error compared to other phase shifters. Low loss and reduction in size of the phased array antenna can be done by fabricating the phase shifter using surface micromachining or bulk micromachining (Topalli et al., 2008). The 3 bit low loss DMTL phase shifter (Topalli et al., 2008) and 5 bit DMTL phase shifters were designed using surface micromachining which is capable of providing better phase resolution for Airborne and Satellite On the Move (SOTM) applications(Pomona et al., 2011). The detailed literature on types of Digital Radio Frequency MEMS phase shifter is briefed (Source: Amrita Chakraborty & Bhaskar Gupta 2017)

## 4.1 Distributed MEMS Transmission Line (DMTL) Phase Shifter

Distributed MEMS Transmission Line phase shifter (Bhosale & Kolhare, 2013) can be achieved by periodic loading of membranes over the Coplanar wave guide transmission lines. DMTL phase shifter operates on the principle of dispersion of waves. When the distributed capacitance on a transmission line increases this in turn will decrease the phase velocity thus providing a differential phase shift. The DMTL phase shifter can be built using CPW and microstrip lines .

Schematic representation of DMTL Phase Shifter is shown in Figure 6 loaded using similar MEMS membrane. The two ends of the membrane are permanently fixed in the ground conductor and the actuation of the membrane is controlled by electrostatic actuation mechanism The loaded MEMS membrane are placed with equal spacing (S), where l is the length of the membrane, w is the width of the membrane, n is the number of MEMS membrane and L is the length of the DMTL phase shifter.

*Figure 6. Schematic of DMTL phase shifter loaded with MEMS membrane*

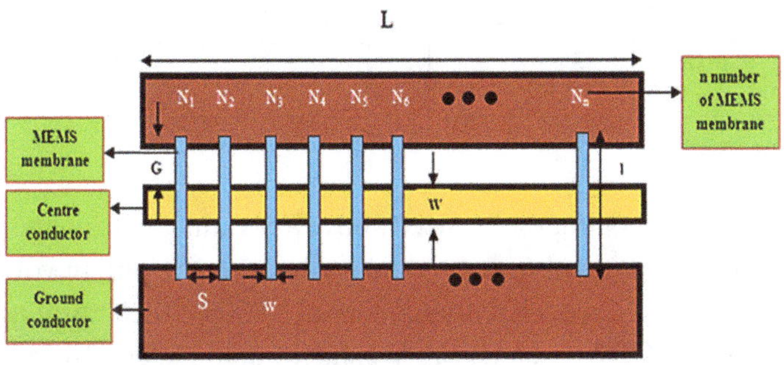


## 5. KEY ISSUES IN RF MEMS DEVICES

Exploiting RF MEMS has been slow due to a number of persistent issues, including affordable packaging, sufficient reliability, low speed, power handling, high actuation voltage, reliability, packaging, cost. Different topologies are currently available to improve the performance of the devices and still many issues need profound concentration during designing, fabrication and analyzing.

### 5.1. Design

In order to develop a electrostatically activated RF MEMS switch, structural modifications of the switch must be performed so as to provide very low OFF state insertion loss, actuation voltage, the highest ON state isolation, and increased switching time. Most of the RF MEMS switches using metal membranes are developed on a CPW platform to achieve size reduction and increased Radio frequency performance (Qian et al., 2000).

Length of the MEMS membrane used for actuation as a switch is determined by the widths and gaps of CPW transmission line . The length of the membrane will have great influence on speed, insertion loss, isolation and pull in voltage of the switch. The constraints present in CPW can be overcome using Elevated Coplanar Waveguide (ECPW). An ECPW switch based on electro static actuation is given in Figure 5.1 (Kanthamani et al., 2009) which has better electrostatic performance than the RF MEMS switches realized on CPW platform.

*Figure 7. RF MEMS shunt switch on ECPW platform*

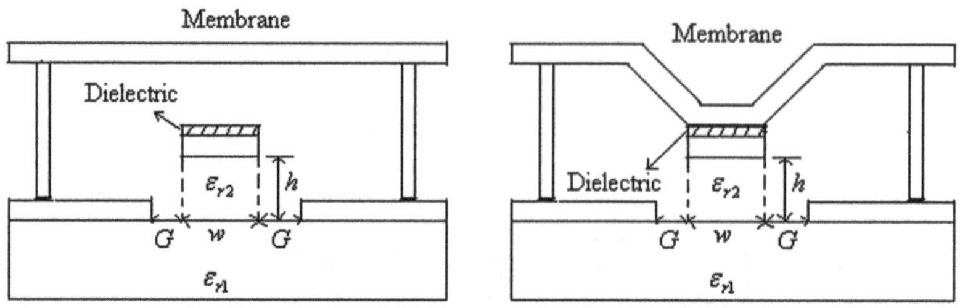


In ECPW the center conductor is raised above a certain level and the gap between the center conductor and membrane gets reduced so that the pull in voltage required for actuation also gets reduced(Kanthamani et al., 2009). Structural modifications could also be done in CPW such as SAW shaped, using space filling curves such as Hilbert, peano, H- order and Moores curves by Amrita Chakraborty & Bhaskar Gupta 2017)

Distributed MEMS Transmission Line (DMTL) phase shifters are by loading MEMS membrane over a CPW transmission line in a linear fashion. The lateral dimension of the phase shifter increases with the degree of phase shift and also increases the overall size of the device and hence occupies more substrate area (Lacroix et al., 2008).The minimum substrate area of 8.39mmx0.49mm occupied by the linear phase shifter is reported in [77].The compact 2 and 3- bit DMTL phase shifter designed with a size of 1.5mm x 2.1mm and 3.2mm x 2.1mm is reported in (Varian & Walton, 2002). Another way of achieving size reduction in DMTL phase shifter is by using Defected Ground Structures (DGS). Size reduction can be done using electromagnetic band gap (Sengupta & Kiran, 2015).Extended size reduction of 23% is achieved by designing the phase shifter using fractals at Ku band (Du et al., 2010; Jahanbakht et al., 2008).

The DMTL phase shifter design could be formulated using space filling curves for further size reduction(Chakraborty et al., 2011).The centre conductor of the CPW transmission line is patterned over the substrate in such a way that it retains the shape of the specified space filling curve to obtain the desired response. The most compact $180^0$ phase shifter is designed using meander lines(Chakraborty et al., 2012), Moore curve (1.5x1.5mm), modified Hilbert curve (0.82x0.82mm), peano curve (1x0.88mm) (Chakraborty & Gupta, 2016a; Chakraborty & Gupta, 2016b; Chakraborty & Gupta, 2017; Chakraborty & Kar, 2016) evidences the size reduction and performance compared to conventional DMTL phase shifter. H - order space filling curve - a modified version of Hilbert approximations has superior locality approximation better than Hilbert indexing curves roughly by 50% (Haverkort & van Walderveen, 2010; Niedermeier et al., 1997).

## 5.2. Analysis

All MEMS structures which are three-dimensional (3-D), are geometrically complex and are electromechanically coupled and hence development of Computer Aided Design (CAD) tools for the analysis of MEMS is more complicated than the traditional mechanical or electrical systems. The analysis of 3-D MEMS involves two coupled domains, namely 3D mechanical analysis and 2D electrostatic analysis as in the literature(Bathe et al., 1975; Nabors & White, 1991; Phillips & White, 1994). Various Coupled domain analysis of MEMS switches such as relaxation methods(Gilbert et al., 1995; Yie et al., 1994), surface Newton method(Bachtold et al., 1995; Yie et al., 1994), coupled Newton method(Aluru & White, 1996; Aluru & White, 1997), multilevel Newton method (Aluru & White, 1996) are reported. Algorithms based on relaxation were used to combine Finite element mesh based elasto static analysis and accelerated boundary element-based electrostatic analysis.

Finite Difference Time Domain (FDTD) formulation was proposed (Tentzeris, 2002) to model the moving parts of MEMS switches. Later adaptive generalization using wavelet transform named as multi resolution time domain technique, used to solve computational issues was proposed (Bushyager & Tentzeris, 2001).Since RF MEMS Switch has both the fixed and movable part, a hybrid methodology combines both of these. Finite Element-Boundary Integration (FE-BI) method is used for analyzing the fixed section and Method of Moments (MOM) for analyzing the movable part of RF-MEMS switches (Wang et al., 2004).This method addresses the large scale variation within a single computational domain.

But in all these methods it is unfortunate that we need to generate a uniform or adaptive volume mesh or 3-D mesh elements on the micro device, for elastic analysis and a surface mesh(2-D meshing) are required to perform electrostatic analysis (Ananthasuresh et al., 1996). In these coupled domain analysis methods, complexity increases because volume meshing has to be compatible with the surface meshing. Usage of proper interpolation function for better convergence is also required.

The complexity of mesh generation in micro device can be overcome using an efficient approach such as point based method or mesh less methods for the modeling and design of MEMS devices(Belytschko et al., 1996).Various point based methods are proposed. Reproducing Kernel Particle Method (RKPM) was proposed to circumvent the problems in smooth particle hydrodynamics method and subsequently applied for the analysis of fixed -fixed and cantilever beams over the ground planes in(Aluru, 1999; Aluru, 2000; Aluru & Li, 2001; Li & Aluru, 2003a; Li & Aluru, 2003b). Radial point interpolation method [RPIM] as a modified version of RKPM was used to model bilayer grapheme nano ribbon(Nivetha & Kanthamani, 2019). Recently QLRKPM was used to model a 3bit phase shifter to obtain the pull in characteristics and hence determine the losses (Mary Sindhuja & Kanthamani, 2019).

## 6. CONCLUSION

In today's 5G communication systems, miniaturization has become a significant factor in the development of microwave integrated circuit components. With the advancement in MEMS/NEMS technology, micro/nano structures are being used. Various structural changes has been studied to achieve miniaturization. An elevated coplanar waveguide and space filling curves can be used to model a RF MEMS switch and comparing to conventional CPW structure confirming the possibility of miniaturization.

## REFERENCES

Admschik, M., Ertl, S., Schmid, P., Gluche, P., Flöter, A., & Kohn, E. (1999, June). Electrostatic diamond micro switch. In *Tech. Digest, 10th Int. Conf. on Solid-State Sensors and Actuators* (pp. 1284-7). Academic Press.

Aluru, N. R. (1999). A reproducing kernel particle method for meshless analysis of microelectromechanical systems. *Computational Mechanics*, *23*(4), 324–338. doi:10.1007004660050413

Aluru, N. R. (2000). A point collocation method based on reproducing kernel approximations. *International Journal for Numerical Methods in Engineering*, *47*(6), 1083–1121. doi:10.1002/(SICI)1097-0207(20000228)47:6<1083::AID-NME816>3.0.CO;2-N

Aluru, N. R., & Li, G. (2001). Finite cloud method: A true meshless technique based on a fixed reproducing kernel approximation. *International Journal for Numerical Methods in Engineering*, *50*(10), 2373–2410. doi:10.1002/nme.124

Aluru, N. R., & White, J. (1996, June). Direct-Newton finite-element/boundary-element technique for micro-electro-mechanical analysis. In *Tech. Digest, Solid-State Sensor and Actuator Workshop* (pp. 54-57). Academic Press.

Aluru, N. R., & White, J. (1997). An efficient numerical technique for electrochemical simulation of complicated microelectromechanical structures. *Sensors and Actuators. A, Physical*, *58*(1), 1–11. doi:10.1016/S0924-4247(97)80218-X

Ananthasuresh, G. K., Gupta, R. K., & Senturia, S. D. (1996). An approach to macromodeling of MEMS for nonlinear dynamic simulation. *Microelectromechanical Systems*, *59*, 401–407.

Bachtold, M., Korvink, J. G., Funk, J., & Baltes, H. (1995, December). New convergence scheme for self-consistent electromechanical analysis of iMEMS. In *Proceedings of International Electron Devices Meeting* (pp. 605-608). IEEE. 10.1109/IEDM.1995.499294

Barker, S., & Rebeiz, G. M. (1998). Distributed MEMS true-time delay phase shifters and wide-band switches. *IEEE Transactions on Microwave Theory and Techniques*, *46*(11), 1881–1890. doi:10.1109/22.734503

Bathe, K. J., Ramm, E., & Wilson, E. L. (1975). Finite element formulations for large deformation dynamic analysis. *International Journal for Numerical Methods in Engineering*, *9*(2), 353–386. doi:10.1002/nme.1620090207

Belytschko, T., Krongauz, Y., Organ, D., Fleming, M., & Krysl, P. (1996). Meshless methods: An overview and recent developments. *Computer Methods in Applied Mechanics and Engineering*, *139*(1), 3–47. doi:10.1016/S0045-7825(96)01078-X

Bhosale, P. S., & Kolhare, N. R. (2013). A review paper on design of distributed RF MEMS phase shifter for aerospace communication. *International Journal of Engineering Trends and Technology*, *4*(3), 284–288.

Brown, E. R. (1999, December). Architectural issues in RF-MEMS based reconfigurable apertures. *NSF RF Micromachining MEMS Technol. Wireless Commun. Syst. Workshop.*

Bushyager, N. A., & Tentzeris, M. M. (2001, May). Modeling and design of RF MEMS structures using computationally efficient numerical techniques. In *2001 Proceedings. 51st Electronic Components and Technology Conference (Cat. No. 01CH37220)* (pp. 327-330). IEEE. 10.1109/ECTC.2001.927744

Cetiner, B. A., Qian, J. Y., Chang, H. P., Bachman, M., Li, G. P., & De Flaviis, F. (2003). Monolithic integration of RF MEMS switches with a diversity antenna on PCB substrate. *IEEE Transactions on Microwave Theory and Techniques*, *51*(1), 332–335. doi:10.1109/TMTT.2002.806521

Chakraborty, A., & Gupta, B. (2016a). Development of compact 180° phase shifters based on MEMS technology. *Sensors and Actuators. A, Physical*, *247*, 187–198. doi:10.1016/j.sna.2016.05.046

Chakraborty, A., & Gupta, B. (2016b). Paradigm phase shift: RF MEMS phase shifters: An overview. *IEEE Microwave Magazine*, *18*(1), 22–41. doi:10.1109/MMM.2016.2616155

Chakraborty, A., & Gupta, B. (2017). Utility of RF MEMS miniature switched capacitors in phase shifting applications. *AEÜ. International Journal of Electronics and Communications*, *75*, 98–107. doi:10.1016/j.aeue.2017.03.011

Chakraborty, A., & Kar, A. K. (2016). Utilization of Fractal Geometry for Phase Shifter Implementation. In *Proceedings of the International Conference on Signal, Networks, Computing, and Systems* (pp. 317-323). Springer. 10.1007/978-81-322-3589-7_35

Chakraborty, A., Kundu, A., Chatterjee, S., & Gupta, B. (2012). *Design of Compact RF MEMS Phase Shifters Based on Space-filling Curves*. Academic Press.

Chakraborty, A., Kundu, A., Dhar, S., Maity, S., Chatterjee, S., & Gupta, B. (2011, September). Compact K-band distributed RF MEMS phase shifter based on high-speed switched capacitors. In *2011 11th Mediterranean Microwave Symposium (MMS)* (pp. 25-28). IEEE. 10.1109/MMS.2011.6068521

De Los Santos, H. (2002). *RF MEMS circuit design for wireless communications*. Artech.

De Los Santos, H. J., Kao, Y. H., Caigoy, A. L., & Ditmars, E. D. (1997, February). *Microwave and mechanical considerations in the design of MEM switches for aerospace applications. In 1997 IEEE Aerospace Conference* (Vol. 3). IEEE.

Dey, S., Koul, S. K., Poddar, A. K., & Rohde, U. L. (2017). Ku to V-band 4-bit MEMS phase shifter bank using high isolation SP4T switches and DMTL structures. *Journal of Micromechanics and Microengineering*, *27*(10), 105010. doi:10.1088/1361-6439/aa8071

Dey, S., Koul, S. K., Poddar, A. K., & Rohde, U. L. (2018). Reliable and compact 3-and 4-bit phase shifters using MEMS SP4T and SP8T switches. *Journal of Microelectromechanical Systems*, *27*(1), 113–124. doi:10.1109/JMEMS.2017.2782780

Du, Y., Bao, J., Wu, W., Tu, C., & Mu, P. (2010, July). Design of a novel 5-bit distributed MEMS phase shifter. In *2010 International Conference on Communications, Circuits and Systems (ICCCAS)* (pp. 706-709). IEEE.

Garver, R. V. (1972). Broad-band diode phase shifters. *IEEE Transactions on Microwave Theory and Techniques*, *20*(5), 314–323. doi:10.1109/TMTT.1972.1127751

Gaspari, R. A., & Yee, H. H. (1978, June). Microwave GaAs FET switching. In *1978 IEEE-MTT-S International Microwave Symposium Digest* (pp. 58-60). IEEE.

Gilbert, J. R., Legtenberg, R., & Senturia, S. D. (1995, February). 3D coupled electro-mechanics for MEMS: applications of CoSolve-EM. In *Proceedings IEEE Micro Electro Mechanical Systems* (p. 122). IEEE. 10.1109/MEMSYS.1995.472542

Glance, B. (1979). A fast low-loss microstrip PIN phase shifter. *IEEE Transactions on Microwave Theory and Techniques*, *27*(1), 14–16. doi:10.1109/TMTT.1979.1129551

Goldsmith, C., Randall, J., Eshelman, S., Lin, T. H., Denniston, D., Chen, S., & Norvell, B. (1996, June). Characteristics of micromachined switches at microwave frequencies. In *1996 IEEE MTT-S International Microwave Symposium Digest. IEEE MTT-S International Microwave Symposium*, *2*, 1141–1144.

Goldsmith, C. L., Yao, Z., Eshelman, S., & Denniston, D. (1998). Performance of low-loss RF MEMS capacitive switches. *IEEE Microwave and Guided Wave Letters*, *8*(8), 269–271. doi:10.1109/75.704410

Gretillat, M. A., Gretillat, F., & De Rooij, N. F. (1999). Micromechanical relay with electrostatic actuation and metallic contacts. *Journal of Micromechanics and Microengineering, 9*(4), 324.

Gretillat, M. A., Gretillat, F., & De Rooij, N. F. (1999). Micromechanical relay with electrostatic actuation and metallic contacts. *Journal of Micromechanics and Microengineering*, *9*(4), 324–331. doi:10.1088/0960-1317/9/4/307

Guclu, C., Cetintepe, C., Civi, O. A., Demir, S., & Akin, T. (2010, August). 35 GHz phased array antenna using DMTL phase shifters. In *2010 10th Mediterranean Microwave Symposium* (pp. 114-117). IEEE.

Haverkort, H., & van Walderveen, F. (2010). Locality and bounding-box quality of two-dimensional space-filling curves. *Computational Geometry*, *43*(2), 131–147. doi:10.1016/j.comgeo.2009.06.002

Hayden, J. S., & Rebeiz, G. M. (2002, June). A low-loss Ka-band distributed MEMS 2-bit phase shifter using metal-air-metal capacitors. In *2002 IEEE MTT-S International Microwave Symposium Digest (Cat. No. 02CH37278)* (Vol. 1, pp. 337-340). IEEE. 10.1109/MWSYM.2002.1011625

Hong, W., Jiang, Z. H., Yu, C., Zhou, J., Chen, P., Yu, Z., ... Cheng, Y. (2017). Multibeam antenna technologies for 5G wireless communications. *IEEE Transactions on Antennas and Propagation*, *65*(12), 6231–6249. doi:10.1109/TAP.2017.2712819

Hosaka, H., Kuwano, H., & Yanagisawa, K. (1994). Electromagnetic microrelays: Concepts and fundamental characteristics. *Sensors and Actuators. A, Physical, 40*(1), 41–47. doi:10.1016/0924-4247(94)85028-3

Hyman, D., Lam, J., Warneke, B., Schmitz, A., Hsu, T. Y., Brown, J., ... Lee, J. (1999). Surface-micromachined RF MEMs switches on GaAs substrates. *International Journal of RF and Microwave Computer-Aided Engineering, 9*(4), 348-361.

Hyman, D., & Mehregany, M. (1998, October). Contact physics of gold microcontacts for MEMS switches. In *Electrical Contacts-1998. Proceedings of the Forty-Fourth IEEE Holm Conference on Electrical Contacts (Cat. No. 98CB36238)* (pp. 133-140). IEEE. 10.1109/HOLM.1998.722438

Iannacci, J. (2017). RF-MEMS for high-performance and widely reconfigurable passive components–A review with focus on future telecommunications, Internet of Things (IoT) and 5G applications. *Journal of King Saud University. Science, 29*(4), 436–443. doi:10.1016/j.jksus.2017.06.011

Iannacci, J., Huhn, M., Tschoban, C., & Pötter, H. (2016a). RF-MEMS technology for 5G: Series and shunt attenuator modules demonstrated up to 110 GHz. *IEEE Electron Device Letters, 37*(10), 1336–1339. doi:10.1109/LED.2016.2604426

Iannacci, J., Huhn, M., Tschoban, C., & Pötter, H. (2016b). RF-MEMS technology for future mobile and high-frequency applications: Reconfigurable 8-bit power attenuator tested up to 110 GHz. *IEEE Electron Device Letters, 37*(12), 1646–1649. doi:10.1109/LED.2016.2623328

Jahanbakht, M., Naser-Moghadasi, M., & Lotfi-Neyestanak, A. A. (2008). Fractal beam KU-band MEMS phase shifter. *Progress In Electromagnetics Research, 5*, 73–85. doi:10.2528/PIERL08101703

Kanthamani, S., Mohan, S., Raju, S., Abhaikumar, V., & Mohan, V. (2009). Meshless Analysis of Radio Frequency Microelectromechanical Systems Shunt Switch. *Defence Science Journal, 59*(6), 622–626. doi:10.14429/dsj.59.1567

Katehi, L. P. (1995). Si micromachining in high-frequency applications. *CRC Handbook on Si Micromachining*, 1771-1787.

Katehi, L. P., Harvey, J. F., & Brown, E. (2002). MEMS and Si micromachined circuits for high-frequency applications. *IEEE Transactions on Microwave Theory and Techniques, 50*(3), 858–866. doi:10.1109/22.989969

Kim, C. S., Lee, C. W., & Borer, J. R. (1966). Varactor S-band direct phase modulator. *IEEE Journal of Solid-State Circuits, 1*(1), 45–51. doi:10.1109/JSSC.1966.1049755

Kim, H. T., Park, J. H., Lee, S., Kim, S., Kim, J. M., Kim, Y. K., & Kwon, Y. (2002). V-band 2-b and 4-b low-loss and low-voltage distributed MEMS digital phase shifter using metal-air-metal capacitors. *IEEE Transactions on Microwave Theory and Techniques, 50*(12), 2918–2923. doi:10.1109/TMTT.2002.805285

Komura, Y., Sakata, M., Seki, T., Kobayashi, K., Sano, K., Horiike, S., & Ozawa, K. (1999, April). Micro machined relay for high frequency application. In Proceedings Of The Relay Conference-National Association Of Relay Manufacturers (Vol. 47, pp. 12-12). Academic Press.

Kruglick, E. J., & Pister, K. S. (1998, June). *Bistable MEMS relays and contact characterization. 1998 Solid-State Sensor and Actuator Workshop.*

Kruglick, E. J., & Pister, K. S. (1999). Lateral MEMS microcontact considerations. *Journal of Microelectromechanical Systems, 8*(3), 264–271. doi:10.1109/84.788630

Lacroix, B., Pothier, A., Crunteanu, A., & Blondy, P. (2008, October). Phase shifter design based on fast RF MEMS switched capacitors. In *2008 European Microwave Integrated Circuit Conference* (pp. 478-481). IEEE. 10.1109/EMICC.2008.4772333

Larson, L. E., Hackett, R. H., & Lohr, R. F. (1991, June). Microactuators for GaAs-based microwave integrated circuits. In *TRANSDUCERS'91: 1991 International Conference on Solid-State Sensors and Actuators. Digest of Technical Papers* (pp. 743-746). IEEE. 10.1109/SENSOR.1991.148990

Li, G., & Aluru, N. R. (2003a). A boundary cloud method with a cloud-by-cloud polynomial basis. *Engineering Analysis with Boundary Elements, 27*(1), 57–71. doi:10.1016/S0955-7997(02)00110-8

Li, G., & Aluru, N. R. (2003b). Efficient mixed-domain analysis of electrostatic MEMS. *IEEE Transactions on Computer-Aided Design of Integrated Circuits and Systems, 22*(9), 1228–1242. doi:10.1109/TCAD.2003.816210

Lim, K., Pinel, S., Davis, M., Sutono, A., Lee, C. H., Heo, D., & Tummala, R. (2002). RF-system-on-package (SOP) for wireless communications. *IEEE Microwave Magazine, 3*(1), 88–99. doi:10.1109/MMW.2002.990700

Lucyszyn, S. (2004). Review of radio frequency microelectromechanical systems technology. *IEE Proceedings. Science Measurement and Technology, 151*(2), 93–103. doi:10.1049/ip-smt:20040405

Majumder, S., McGruer, N. E., Zavracky, P. M., Adams, G. G., Morrison, R. H., & Krim, J. (1997, June). Measurement and modelling of surface micromachined, electrostatically actuated microswitches. In *Proceedings of International Solid State Sensors and Actuators Conference (Transducers' 97)* (Vol. 2, pp. 1145-1148). IEEE. 10.1109/SENSOR.1997.635407

Mary Sindhuja, N. M., & Kanthamani, S. (2019). Three-Bit DMTL Phase Shifter for Phased Array Antennas. *Journal of Circuits, Systems, and Computers*, *28*(07), 1950112. doi:10.1142/S0218126619501123

McLevige, W. V., & Sokolov, V. (1980). Microwave switching with parallel-resonated GaAs FETs. *IEEE Electron Device Letters*, *1*(8), 156–158. doi:10.1109/EDL.1980.25271

Miao, M., Wu, G., Li, Z., Jin, Y., & Hao, Y. (2006, January). A bulk micromachined distributed digital microwave phase shifter with butterfly multilayer bridges and MAM capacitors. In *2006 1st IEEE International Conference on Nano/Micro Engineered and Molecular Systems* (pp. 479-483). IEEE. 10.1109/NEMS.2006.334804

Muldavin, J. B., & Rebeiz, G. M. (1999, June). 30 GHz tuned MEMS switches. In *1999 IEEE MTT-S International Microwave Symposium Digest (Cat. No. 99CH36282)* (Vol. 4, pp. 1511-1514). IEEE. 10.1109/MWSYM.1999.780241

Nabors, K., & White, J. (1991). FastCap: A multipole accelerated 3-D capacitance extraction program. *IEEE Transactions on Computer-Aided Design of Integrated Circuits and Systems*, *10*(11), 1447–1459. doi:10.1109/43.97624

Nguyen, C. C., Katehi, L. P., & Rebeiz, G. M. (1998). Micromachined devices for wireless communications. *Proceedings of the IEEE*, *86*(8), 1756–1768. doi:10.1109/5.704281

Niedermeier, R., Reinhardt, K., & Sanders, P. (1997, September). Towards optimal locality in mesh-indexings. In *International Symposium on Fundamentals of Computation Theory* (pp. 364-375). Springer. 10.1007/BFb0036198

Nivetha, J., & Kanthamani, S. (2019). Radial Point Interpolation Method (RPIM) for the Meshless Analysis of Graphene Nano Ribbon. *Journal of the Institution of Electronics and Telecommunication Engineers*, 1–6. doi:10.1080/03772063.2018.1553639

Peroulis, D. (2003). *RF MEMS devices for multifunctional integrated circuits and antennas* (Doctoral dissertation). University of Michigan.

Petersen, K. E. (1979). Micromechanical membrane switches on silicon. *IBM Journal of Research and Development*, *23*(4), 376–385. doi:10.1147/rd.234.0376

Phillips, J. R., & White, J. (1994, November). A precorrected-FFT method for capacitance extraction of complicated 3-D structures. In ICCAD (Vol. 94, pp. 268-271). Academic Press.

Pomona, I., Farinelli, P., Microtech, R. F., Margesin, B., & Meneghesso, G. (2011). RF-MEMS Phase Shifters for Phased Array Antennas. In Nanotech Italy 2011 (No. 13, pp. 37-39). ITA.

Pranonsatit, S., Holmes, A. S., Robertson, I. D., & Lucyszyn, S. (2006). Single-pole eight-throw RF MEMS rotary switch. *Journal of Microelectromechanical Systems*, *15*(6), 1735–1744. doi:10.1109/JMEMS.2006.883578

Qian, J. Y., Li, G. P., & De Flavis, F. (2000, June). A parametric model of MEMS capacitive switch operating at microwave frequencies. In *2000 IEEE MTT-S International Microwave Symposium Digest (Cat. No. 00CH37017)* (Vol. 2, pp. 1229-1232). IEEE. 10.1109/MWSYM.2000.863581

Real, S. M., Turjo, R. R., Sobhan, F. F., & Ashraf, N. (2018). *Phased array antennas for mmWave frequency spectrum for 5G communication* (Doctoral dissertation). BRAC University.

Rebeiz, G. M. (2004). *RF MEMS: theory, design, and technology*. John Wiley & Sons.

Rebeiz, G. M., & Muldavin, J. B. (2001). RF MEMS switches and switch circuits. *IEEE Microwave Magazine*, *2*(4), 59–71. doi:10.1109/6668.969936

Rebeiz, G. M., Tan, G. L., & Hayden, J. S. (2002). RF MEMS phase shifters: Design and applications. *IEEE Microwave Magazine*, *3*(2), 72–81. doi:10.1109/MMW.2002.1004054

Sanders, C. (1998). *MCNC thermally actuated microrelays*. News Release.

Schiele, I., & Hillerich, B. (1999). Comparison of lateral and vertical switches for application as microrelays. *Journal of Micromechanics and Microengineering*, *9*(2), 146–150. doi:10.1088/0960-1317/9/2/310

Schiffman, B. M. (1958). A new class of broad-band microwave 90-degree phase shifters. *I.R.E. Transactions on Microwave Theory and Techniques*, *6*(2), 232–237. doi:10.1109/TMTT.1958.1124543

Schlaak, H. F. (1997). Silicon-microrelay-a small signal relay with electrostatic actuator. *Proc. 45th relay conf.*

Sengupta, S., & Kiran, K. U. (2015). Miniatured phase shifter. *Indian Journal of Science and Technology*, *8*(19), 1–7. doi:10.17485/ijst/2015/v8i19/76867

Shekhar, S., Vinoy, K. J., & Ananthasuresh, G. K. (2018). Low-voltage high-reliability MEMS switch for millimeter wave 5G applications. *Journal of Micromechanics and Microengineering*, *28*(7), 075012. doi:10.1088/1361-6439/aaba3e

Simon, J., Saffer, S., & Kim, C. J. (1996, February). A micromechanical relay with a thermally-driven mercury micro-drop. In *Proceedings of Ninth International Workshop on Micro Electromechanical Systems* (pp. 515-520). IEEE. 10.1109/MEMSYS.1996.494035

Sovero, E. A., Mihailovich, R., Deakin, D. S., Higgins, J. A., Yao, J. J., DeNatale, J. F., & Hong, J. H. (1999, August). Monolithic GaAs PHEMT MMICs integrated with high performance MEMS microrelays. In *1999 SBMO/IEEE MTT-S International Microwave and Optoelectronics Conference* (Vol. 1, pp. 257-260). IEEE. 10.1109/IMOC.1999.867103

Sun, X. Q., Farmer, K. R., & Carr, W. N. (1998, January). A bistable microrelay based on two-segment multimorph cantilever actuators. In *Proceedings MEMS 98. IEEE. Eleventh Annual International Workshop on Micro Electro Mechanical Systems. An Investigation of Micro Structures, Sensors, Actuators, Machines and Systems (Cat. No. 98CH36176* (pp. 154-159). IEEE.

Suzuki, K., Chen, S., Marumoto, T., Ara, Y., & Iwata, R. (1999, June). A micromachined RF microswitch applicable to phased-array antennas. In *1999 IEEE MTT-S International Microwave Symposium Digest (Cat. No. 99CH36282)* (Vol. 4, pp. 1923-1926). IEEE. 10.1109/MWSYM.1999.780350

Taylor, W. P., Allen, M. G., & Dauwalter, C. R. (1997). Batch fabricated electromagnetic microrelays. In Proceedings of the relay conference-national association of relay manufacturers (vol. 45, pp. 8-1). National Association of Relay Manufacturers.

Tentzeris, E. M. (2002, October). Adaptive modeling and design of highly integrated 3D microwave-millimeter wave radio front-ends. In 24th Annual Technical Digest Gallium Arsenide Integrated Circuit (GaAs IC) Symposium (pp. 109-112). IEEE.

Tilmans, H. A., Fullin, E., Ziad, H., Van de Peer, M. D., Kesters, J., Van Geffen, E., . . . Naso, F. (1999, January). A fully-packaged electromagnetic microrelay. In *Technical Digest. IEEE International MEMS 99 Conference. Twelfth IEEE International Conference on Micro Electro Mechanical Systems (Cat. No. 99CH36291)* (pp. 25-30). IEEE. 10.1109/MEMSYS.1999.746746

Topalli, K., Civi, Ö. A., Demir, S., Koc, S., & Akin, T. (2008). A monolithic phased array using 3-bit distributed RF MEMS phase shifters. *IEEE Transactions on Microwave Theory and Techniques*, *56*(2), 270–277. doi:10.1109/TMTT.2007.914377

Tsoulos, G. V. (2000). *Adaptive antennas for wireless communications*. Wiley-IEEE Press.

Varadan, V. K., Vinoy, K. J., & Jose, K. A. (2003). *RF MEMS and their Applications. Pennsylvania State University*. Wiley Online Library.

Varadan, V. K., Vinoy, K. J., Jose, K. A., & Mems, R. (2003). *Their Applications*. Pennsylvania State University.

Varian, K., & Walton, D. (2002). A 2-bit RF MEMS phase shifter in a thick-film BGA ceramic package. *IEEE Microwave and Wireless Components Letters*, *12*(9), 321–323. doi:10.1109/LMWC.2002.803201

Wang, J., Ren, Z., & Nguyen, C. C. (2004). 1.156-GHz self-aligned vibrating micromechanical disk resonator. *IEEE Transactions on Ultrasonics, Ferroelectrics, and Frequency Control*, *51*(12), 1607–1628. doi:10.1109/TUFFC.2004.1386679 PMID:15690722

White, J. F. (1974). Diode phase shifters for array antennas. *IEEE Transactions on Microwave Theory and Techniques*, *22*(6), 658–674. doi:10.1109/TMTT.1974.1128309

Wright, J. A., & Tai, Y. C. (1999, January). Magnetostatic MEMS relays for the miniaturization of brushless DC motor controllers. In *Technical Digest. IEEE International MEMS 99 Conference. Twelfth IEEE International Conference on Micro Electro Mechanical Systems (Cat. No. 99CH36291)* (pp. 594-599). IEEE. 10.1109/MEMSYS.1999.746895

Yao, J. J., & Chang, M. C. (1995, December). Surface micromachined miniature switch for telecommunications applications with signal frequencies from DC up to 4 GHZ. *Proceedings of the 1995 8th International Conference on Solid-State Sensors and Actuators and Eurosensors IX. Part 1 (of 2)*. 10.1109/SENSOR.1995.721827

Yao, Z. J., Chen, S., Eshelman, S., Denniston, D., & Goldsmith, C. (1999). Micromachined low-loss microwave switches. *Journal of Microelectromechanical Systems*, *8*(2), 129–134. doi:10.1109/84.767108

Yie, H., Cai, X., & White, J. (1994, June). Convergence properties of relaxation versus the surface-Newton generalized-conjugate residual algorithm for self-consistent electromechanical analysis of 3-D micro-electro-mechanical structures. In *Proceedings of International Workshop on Numerical Modeling of processes and Devices for Integrated Circuits: NUPAD V* (pp. 137-140). IEEE. 10.1109/NUPAD.1994.343472

# Chapter 2
# Wearable Reconfigurable Antenna

**Reji V**
*SRM Institute of Science and Technology, India*

**C. T. Manimegalai**
*SRM Institute of Science and Technology, India*

## ABSTRACT

*Printed antennas are designed for various applications in recent years. The latest development of antennas are wearable antennas; these wearable antennas are playing a major role in the energy harvesting mechanisms, patient monitoring systems, and military applications. In this chapter a reconfigurable wearable antenna is proposed. Reconfigurable wearable antennas are able to operate in both modes on body and off body. These antennas are reconfigured from one frequency to another frequency depending upon the requirement. Three types of reconfigurable antenna modes are designed and analyzed in the chapter. Frequency reconfiguration with off body mode and on body mode, frequency reconfiguration between inside receiver and outside interrogator, and frequency reconfiguration with different substrate conditions. The choice of frequency bands are WBAN and other wireless application bands. Different switches are used to control the frequencies.*

## INTRODUCTION

Wearable antennas are the major parts of the wireless body area networks and medical applications. Multiple frequency bands are allocated for the wearable antenna from RF to millimeter range. Wearable antennas are mostly developed to wear outside of

DOI: 10.4018/978-1-7998-7611-3.ch002

the body on wrists and dresses. In the wrist it is integrated with watches and wearing bands; But in dresses, this type of antennas is integrated with buttons, belt hooks, and printed in cloths. These wearable antennas are attached or implanted with bodies like a sensor and these are interconnected with each other. The communications between these types of antenna sensors are performed through a gateway. Some antennas are communicated with a master node in ON body communication and it acts as a transceiver for OFF body communication. New development wearable antennas are implanted directly inside the human body with like medical devices (pacemaker, endoscopic machine) to monitor the heartbeat, blood pressure, and sugar levels of the body. Sensor type antennas are implanted inside the head for monitoring various parameters of the human head. The allotted frequency range for body area applications is 401 to 406MHz for medical implant communication service,2.4-5GHz for Wi-Fi small hospital beds and mobile nursing stations,2.4GHz, and 5.8GHz frequencies are used for ISM bands 3.1 to 10.76GHz used for UWB medical services. Frequency bands 59 to 66GHz are allotted for millimeter-wave wearable antenna applications.

## BACKGROUND

Wearable antennas are classified as fabric-based antennas; thin substrate polymer-based antennas, various alloy compound microfluidic antennas, and variable printing antennas such as inkjet printing, screen printing, and 3D printing antennas. Fabric-based antennas are mostly developed by printing antennas like embroidery on the garments and saris. These types of antennas are used for military applications. These embroideries are designed by conductive threads. Polymer-based antennas are developed by low permittivity materials. These types of antenna thickness are greater than fabric-based techniques. So these antennas are integrated with buttons and hooks. Mostly PDMS and TC5005 types of substrates materials are used to develop this type of antennas. Next are microfluid techniques, to design this antenna 1mm wide microfluid materials are used. To develop a printing based antenna a silver or gold nanoparticle-based conductive inks are used. RF energy harvesting antennas are also developed; it can be worn in cloth belts using conductive fabrics to collect the energy from the body to the antenna (Rectenna). More researches are going on for designing the wearable antennas; in the future, these wearable antennas are integrated directly with human bodies like embroidery design, where the skin will be act as the substrate.

In this chapter, the reconfigurable mechanism of wearable antennas are discussed. A reconfigurable antenna has been printed on conductive (Simorangkir et al., 2018) polydimethylsiloxane substrate. A small inner layer was drafted on the substrate; all the lumped elements are encapsulated in the inner layers. It shows a narrow frequency reconfigurable range from 2.3GHz to 2.68GHz.Another antenna was designed for ON and OFF body mode (Tong et al., 2018). These reconfigurable antennas are designed for the 2.45 GHz ISM band and the two modes(Masood et al., 2007)[4](Hall & Hao, 2006) are controlled by PIN diode switch with DC bias. A textile antenna was designed for body area network and ISM band (Liu et al., 2012), this antenna is integrated with buttons and the frequencies are controlled by varactor diode at 2.45GHz and 5.8GHz (Hall & Hao, 2006). A dual-band antenna was designed to operate at 2.45GHz and 5.8GHz (Liu & Guo, 2013) for polarization diversity. Four slots are used here along with a magnetic dipole to divert the polarization of the antenna. A dual (Zhu et al., 2016)[10] band antenna has been demonstrated for the ISM band with low radiation parameters. Wearable antennas have been demonstrated (Zhu & Guo, 2016)[15] for a single frequency, dual(Xiaomu et al., 2007)[13], and multiple frequency applications with acceptable gain. A dual-band antenna was designed with two ports for pattern diversity (Xiaomu et al., 2007).

## MAIN FOCUS OF THE CHAPTER

The main focus of the chapter is to design and analysis of reconfigurable antenna for wearable applications under various switching conditions. This antenna design gives multiple frequency radiation for different wireless applications along with WBAN application.

## ANTENNA DESIGN

### The Inverted Reconfigurable V Antenna Design

Inverted V antennas are one of the traveling wave antennas. It is used for high-frequency applications. This antenna has a common ground with two small terminating posts. The terminating resistance of the antenna is 400Ω and all other parameters of the antennas are adjusted for making microstrip frequency range. The legs of the antennas are $8\lambda/2$ at the highest frequency of $4\lambda/2$ at the lowest frequency. The angles between the legs are known as tilting angle, this angle should be greater than 60 degree for getting maximum radiation.

The proposed inverted antenna is designed on the "FR4 epoxy substrate with width 5 and length 6.2cm. The substrate thickness is 0.16cm. Common ground is allotted on the bottom of the substrate. The width and length of the ground are 5 and 1cm. Four legs are used in the inverted antenna design. Outside leg lengths and widths are 4.5cm and 0.8cm. The inner side leg lengths are 3.8 and 0.5cm.The outside and inside post dimensions are 0.3, 0.5cm and 0.2, 0.5cm.An impedance matching section is connected at the end of the antenna along with the PCB end connector(Masood et al., 2007)[4]. This inverted V antenna is designed for wireless body area networks and other wireless frequency applications. This antenna is a dual-band antenna; it can be radiated two different frequencies simultaneously and reconfigured these frequencies to other frequency ranges depend upon their switching conditions. Reconfigurable antennas are reconfigured their frequencies from one frequency to another frequency without changing their original structure of the antenna by using switches. Many switches can be used for reconfiguring purposes such as PIN diode, varactor diode, MEMS switches, and optical switches (Tong et al., 2018)[1]. In this model, PIN diodes are acted us a switch to reconfigure the frequency of the antenna. These PIN diodes are connected between the two legs of the antenna. The length and width of the switches are selected from SMDT switch specification (Operating frequency of the switch from 2 to 6GHz). The overall dimensions of the antenna are listed in the table1. Two sides capacitive line, bias line, and bias pad dimensions also given in the table1.The simulated antenna is shown in fig1. The leg(dipole) lengths, widths and the matching impedances of the antenna are calculated by the following equations,

(i) If $L=\lambda/2$ (1)

$L/d=\text{infinity}$

$Z= (73+j42.5)\Omega$ (2)

(ii) If $L<=\lambda/2$

$L/d=100$

$Z=Rs=67\Omega$

(iii) If $L= 0.925\lambda$

$Z=50\Omega$

L=Length of the dipole

d=Width of the dipole

The capacitive and inductive lengths of the switches are calculated by the following equations

## Shunt L

If l=λ/4

$\omega_o L = X_L = Zo\tan(2\pi l/\lambda_g)$ (3)

## Shunt C

l=λ/4

$(1/\omega_o C) = Xc = Z_o\cot(2\pi l/\lambda_g)$ (4)

## Series L

l=λ/4

$\omega_o L = X_L = Z1\tan(2\pi l/\lambda_g)$ (5)

## Series C

l=λ/4

$X_c = jZ1\tan((2\pi l/\lambda_g)$ (6)

Zo=Terminating Impedance

l=Inductive and Capacitive line length

$\lambda_g$=guide wavelength

$\lambda$=signal wavelength

Z1=Impedance of microstrip line

$\omega_o$=resonant frequency

*Figure 1. Block diagram of reconfigurable antenna*

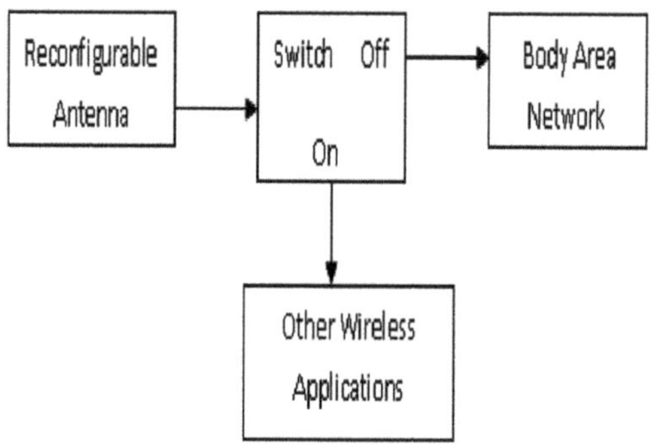


*Table 1. Dimensions of inverted V antenna design*

| Parameter | Dimension (cm) | Parameter | Dimension (cm) |
|---|---|---|---|
| L | 5 | B4 | 0.5 |
| W | 6.2 | B5 | 0.1 |
| P1 | 0.3 | B6 | 0.8 |
| P2 | 0.2 | B7 | 0.62 |
| P3 | 1.5 | W1 | 0.3 |
| P4 | 0.5 | W2 | 0.1 |
| P5 | 0.8 | W3 | 0.2 |
| D1 | 0.3 | W4 | 0.5 |
| B1 | 0.3 | W5 | 0.8 |
| B2 | 0.3 | W6 | 1 |
| B3 | 0.5 | | |

*Figure 2. Reconfigurable antenna dimensions and their electric field pattern*

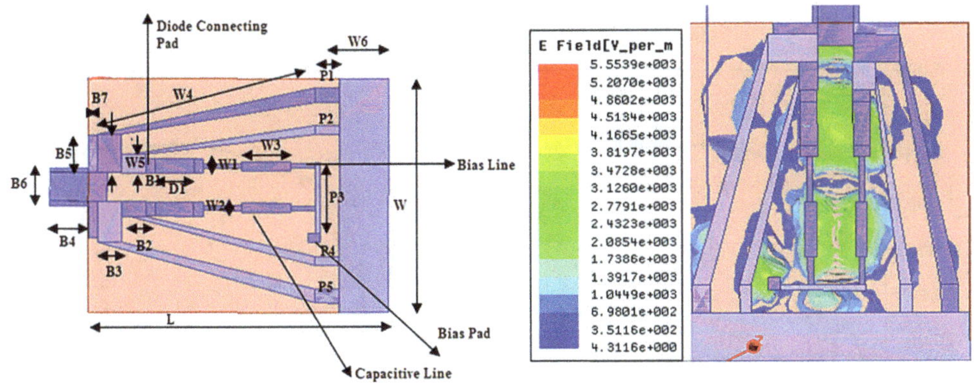


Two PIN diodes are connected to the antenna for switching purposes. The state of the switch is changed by its biasing conditions. Under forward biasing the switches are in ON condition and under the reverse biasing condition the switches are in OFF condition (Hall & Hao, 2006). Four switching logic states are created through the two switches such as 00, 01, 10, and 11. When the switches are in ON condition an equivalent series resistance is connected with inductance across the switch. When the switches are in OFF condition an equivalent capacitance is also connected with resistance across the switch. The simulated equivalent resistance, capacitance, and inductance values are 4Ω, 0.1pf, and 2nH. The equivalent circuits of the switches and biasing circuits are shown in fig.3.

*Figure 3. (a)Biasing Circuit of PIN diode (b)ON state and OFF state of the PIN diode*

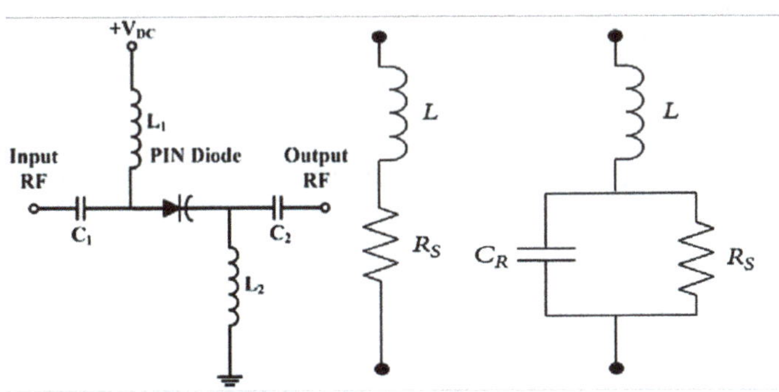

*Figure 4. Reflection co-efficient of antenna under offbody mode conditions for all switching states*

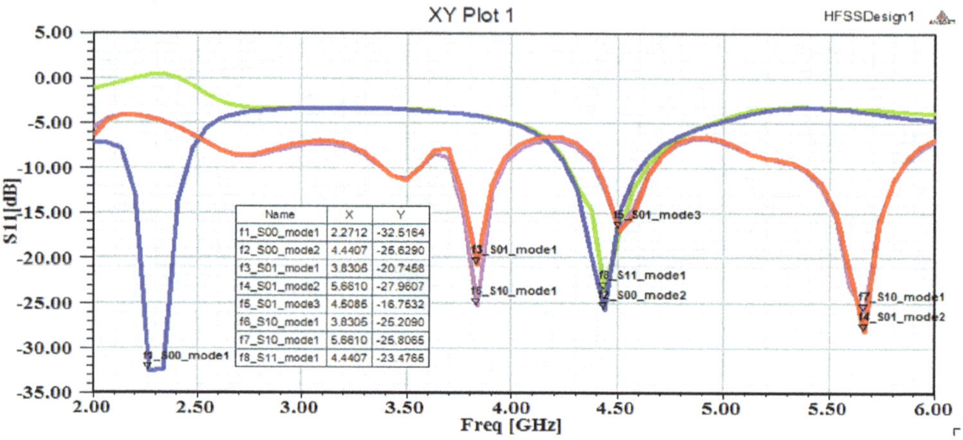


## Antenna Analysis Under OFF Body Mode

The designed reconfigurable antenna is first simulated for OFF body mode (antenna is not attached with the human body) by high frequency simulation software. This reconfigurable antenna is designed for providing dual mode in each switching conditions of the antenna. Two switches are connected between the four legs of the antenna so it provides four switching conditions (Liu et al., 2014)[11]. All switching conditions and there operating frequencies are shown in fig 4.At 00 switching condition the operating frequency of the antennas are 2.2-2.4GHz and 4.4GHz with reflection co-efficient -32dB and -25dB.At 01 condition the antenna is reconfigured the frequencies to 3.8GHz and 5.6GHz with reflection parameter value -20.7dB and -27dB.At 10 and 11 condition of the switch the antenna frequencies are reconfigured to 3.8GHz, 5.6GHz,4.4GHz with acceptable reflection co-efficient value -25dB,-25dB,-23dB.Figure 5 shows the radiation patterns of the antenna for all switching conditios,The φ component of electric field is measured as a function of θ in XY plane when $\theta=0^{o}$ and the θ component of electric field is measured as a function of φ in XZ plane when $\varphi=0^{o}$ .The simulated results shows the maximum eclectic field radiation of the antenna along the $90^{o}$ position.

## Antenna Analysis Under ON Body Mode

There are three mode conditions are considered here (i) Antenna on the Body Tissue(ii) Antenna Analysis Under 1mm inside the Skin (iii) Antenna Analysis under 0.5mm Away From the Body Skin.

*Figure 5. Radiation pattern of antenna under off body mode conditions for 00,01,10,11 switching states*

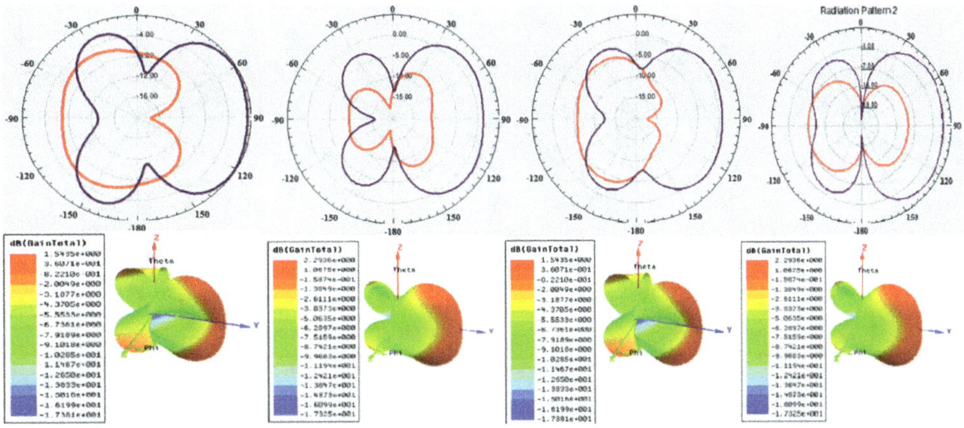

*Figure 6. Antenna on body tissue*

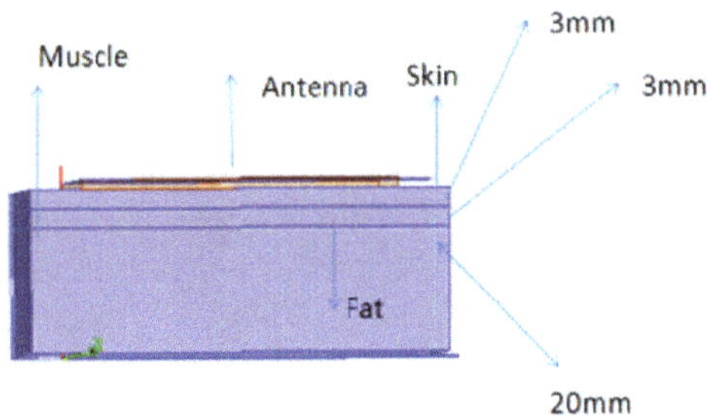


*Table 2. Parameter Values of Human Body Tissue*

| Body Tissue | Conductivity σ | Relative Permittivity $\varepsilon_r$ | Loss Tangent δ |
|---|---|---|---|
| Skin | 1.60 | 42.8 | 0.27 |
| Fat | 0.11 | 5.3 | 0.14 |
| Muscle | 1.74 | 52.7 | 0.24 |

*Figure 7. Reflection co-efficient of antenna on the body tissue for all switching states*

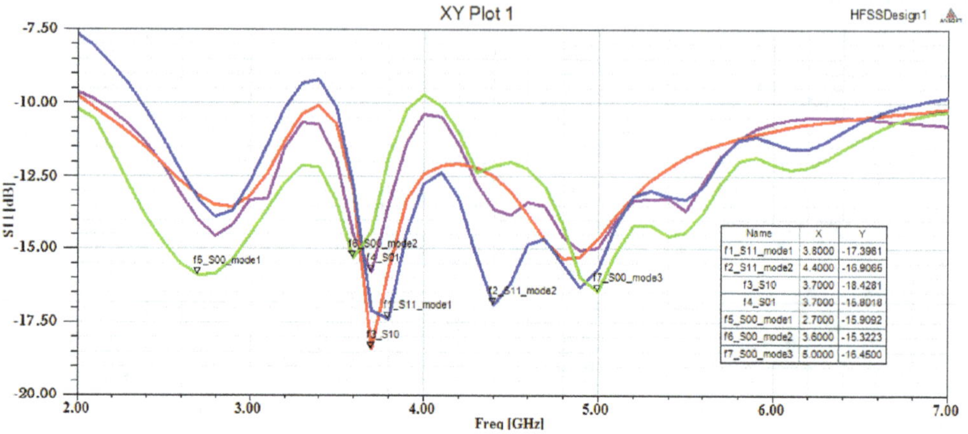


*Figure 8. Radiation pattern of antenna on the body tissue for 00,01,10,11 switching states*

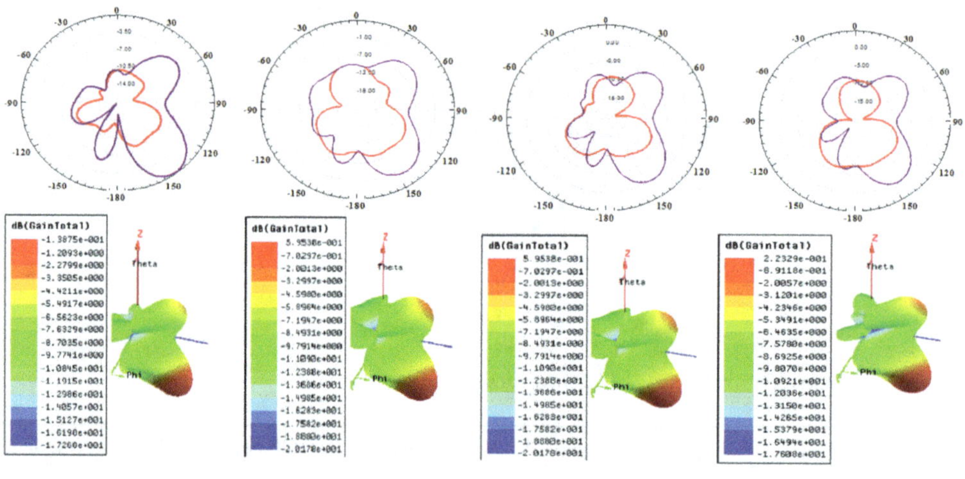

## Antenna on the Body Tissue

The antenna on the body tissue condition the antenna is only touched with the human body tissue not implanted. Under this condition, the antenna can be worn along with wristwatches,(Xiaomu et al., 2007) inside of the belts and inner packets. In this proposed method a tissue box is created by HFSS simulation software to analyze the antenna parameters as shown in figure 6. This tissue box is created for skin, fat, and muscle, The conductivity, relative permittivity, and loss tangent values are listed in the table2. The simulated reflection coefficient and radiation patterns are shown in figure 7,8 for all switching conditions.

The simulated radiating frequencies are 2.7GHz, 3.7GHz, 3.8GHz and 5GHz with good reflection coefficient values. Maximum radiation of the antenna is absorbed by human body tissue along the bottom direction. The radiation patterns of the antenna are shifted towards the 150° (left hand side of the human body) angle direction from 90°.

*Figure 9. Reflection co-efficient of the antenna for all switching condition (1mm inside the skin)*

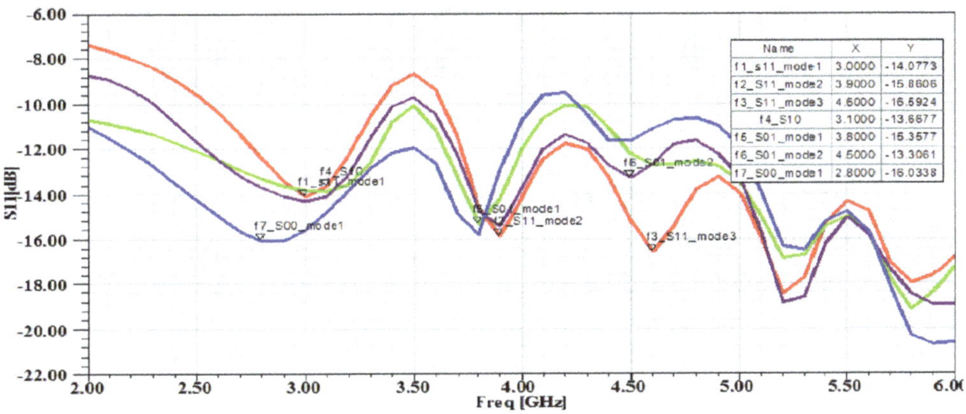


## Antenna Analysis Under 1mm Inside the Skin

In this condition antenna is implanted under 1mm skin depth. The same human body tissue model is considered to analyze the antenna parameters(Antoskiewicz & Shafai, 2017)[15]. The simulated reflection coefficient values are shown in figure 9.Four switching conditions are simulated under this condition. The generated radiating frequencies are 3.9GHz, 4.6GHz,3.8GHz,2.8GHz with good reflection co-efficient values. Figure10 shows the radiation pattern of the antennas for the four switching conditions .Maximum radiation of the antenna is shifted from 90° to 150° angle position .The minor lobe size is also increased in the four switching conditions because of absorption.

*Figure 10. Radiation pattern of the antenna(1mm inside the skin) for 00,01,10,11 switching states*

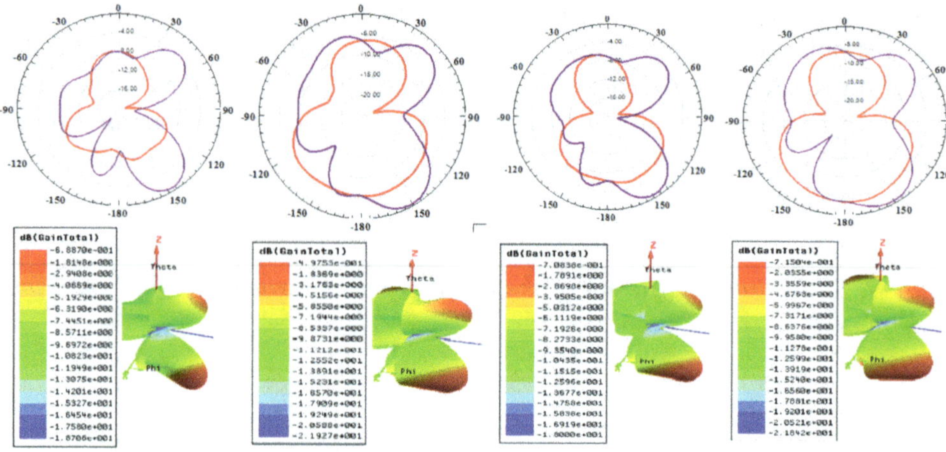

*Figure 11. Reflection Co-efficient of the antenna(0.5mm away from the body) for 00,01,10,11 switching states*

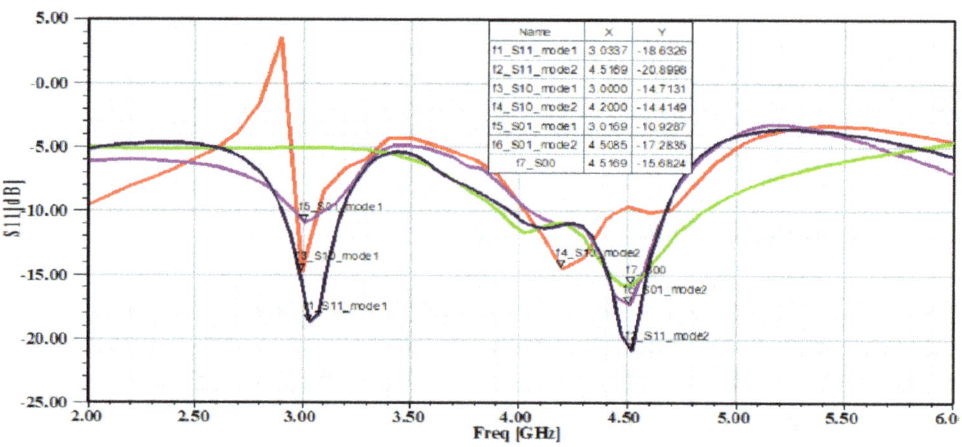


## Antenna Analysis Under 0.5mm Away From the Body Skin

Under this condition antenna is not attached with the body, It can be wore(with helmet or any devices like mobile phone,Walki Talki) with 0.5mm away from the body(Xiaomu et al., 2007). In this condition the following modes appear in different switching conditions.4.5GHz is the dominant mode in this condition as shown in figure11.Fig 12 shows the radiation pattern of the antenna. The maximum radiation of the antenna is along150° and 60° direction.

*Figure 12. Radiation Pattern of the antenna(0.5mm away from the body) for 00,01,10,11 switching states*

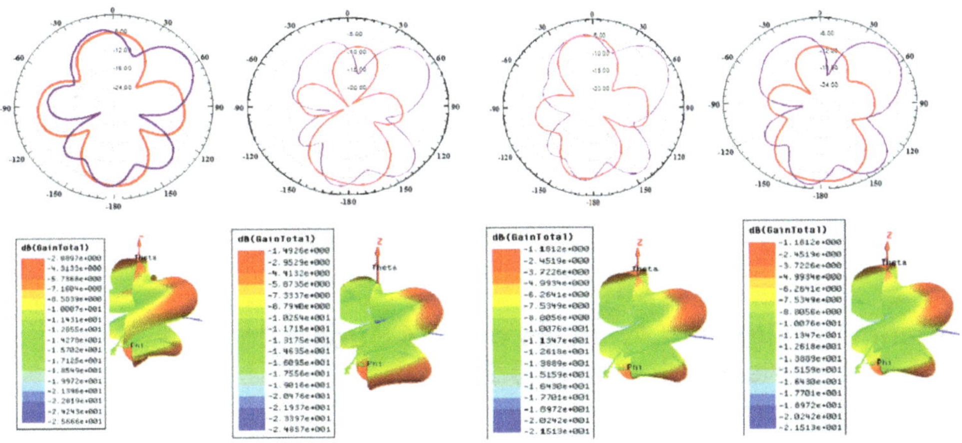

*Figure 13. Reflection Co-efficient of the antenna with substrate thickness h=0.22mm for 00,01,10,11 switching states*

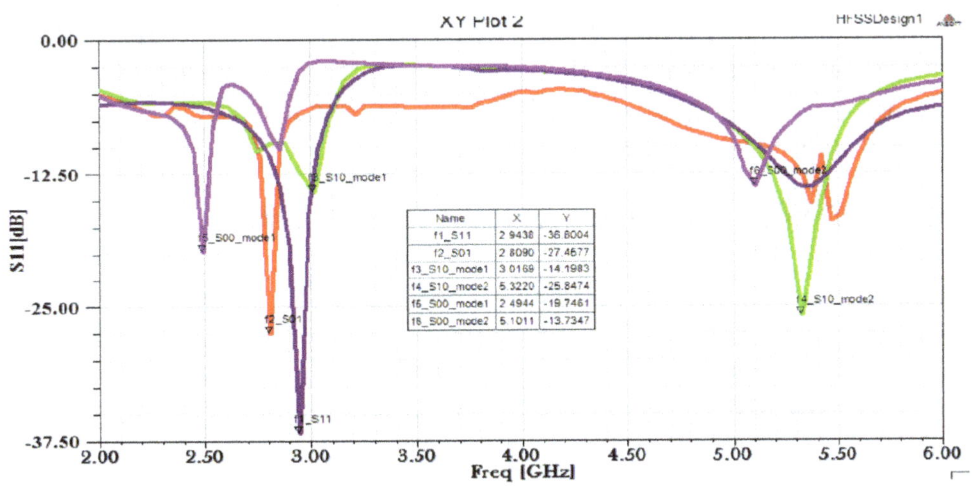

*Figure 14. Radiation Pattern of the antenna with substrate thickness h=0.22mm for 00,01,10,11 switching states*

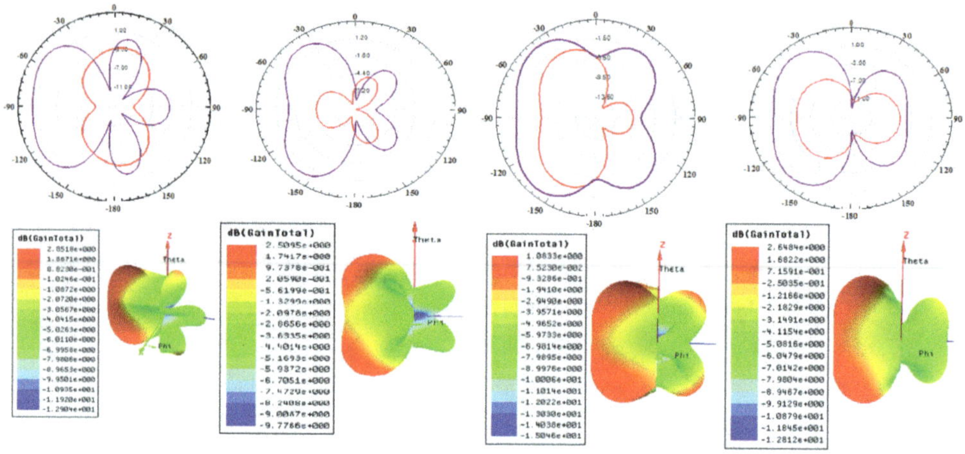

## Antenna Design and Analysis on FR4 Substrate With Substrate Thickness h=0.22mm

The inverted reconfigurable V antenna is also designed on the FR4 substrate with the thickness of the substrate h=0.22mm. This antenna is very thin and easily can be worn with cloths. This antenna is also stitched along with clothes because the substrate of the antenna can be folded. The simulated S11 parameters and radiation patterns are shown in figure13 and 14. These antennas are also used to communicate with the implantable antenna which is connected inside the human body.(Zhu & Guo, 2016) Four switching condition is simulated here, at every switch conditions dual modes are generated from 2.5GHz to 3GHz and 5 to 5.5 GHz. The useful radiating frequencies are 5.4GHz, 2.6GHz, 2.7GHz, 3GHz, 5.38GHz, 5.32GHz with their reflection coefficient values are less than -15dB.The maximum radiation of the antenna is diverted along the -90° position to communicate easily with the implanted human body antennas.

## Antenna Design on Polyester and Denim Substrate

The above inverted V antenna substrate is modified with polyester and denim material to analyze the antenna parameters. The polyester material is selected for the substrate height h=0.22mm and denim material is selected(Shakib & Moghavvemi, 2017) for the substrate height h=1.6m.The reflection co-efficient values of the antennas are shown in figure 15.It does not give good radiation for 2GHz and 6GHz frequency ranges.

*Figure 15. Reflection Co-efficient of the antenna on Polyester(black colour)and Denim(red colour) substrate*

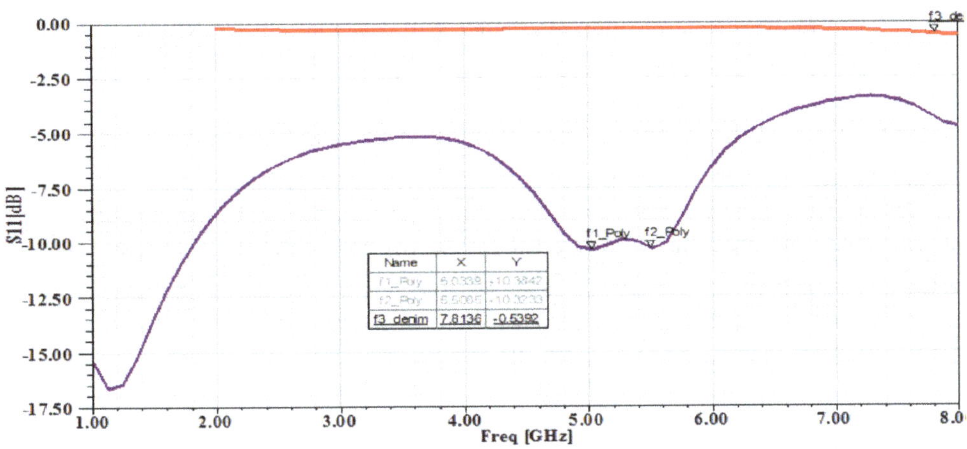


*Table 3. Dimensions of implanted antenna*

| Parameter | Dimension (cm) |
|---|---|
| L | 2.2 |
| W | 1.2 |
| w1 | 0.4 |
| w2 | 0.055 |
| w3 | 0.183 |
| w4 | 0.42 |
| l1 | 1.4 |
| l2 | 1.75 |
| l3 | 0.25 |

## Frequency Reconfiguration Between Inside Receiver and Outside Interrogator

Under these conditions, two antennas are necessary to communicate with each other. One antenna is an interrogator connected outside of the body and another one is a receiver implanted inside the body. If any medical low dimension wearable antenna is connected in the body at 2.4GHz this can be communicated with the outside interrogator antenna. The above antenna(shown in fig 13)acts as the interrogator antenna because it gives the radiation pattern along -90o direction. It is useful to create the radiation along the bottom direction (Simorangkir et al., 2018)[2]. This antenna is designed to change the feeding from the right to the left side to shift the radiation pattern to communicate easily with the inside receiver. The receiver receives the signal from the -90° position of the outside interrogator. The implanted antenna radiation direction should be in any one direction.

The receiver is designed for 2.45Ghz with very low dimensions. The dimensions of the antennas are listed in table 3. It is a low profile less weight antenna. the length and width of the antenna are 1.2 and 2.2cm. The antenna is drafted on the Roger substrate with a substrate thickness of 0.034cm. The simulated antenna and its radiation characteristics are shown in fig16. The antenna is simulated for 2.45GHz with a return loss -20dB and radiation of the antenna along 0o and 180o direction. This antenna can be implanted under the skin.

*Figure 16. Simulated implanted antenna and their radiation characteristics*

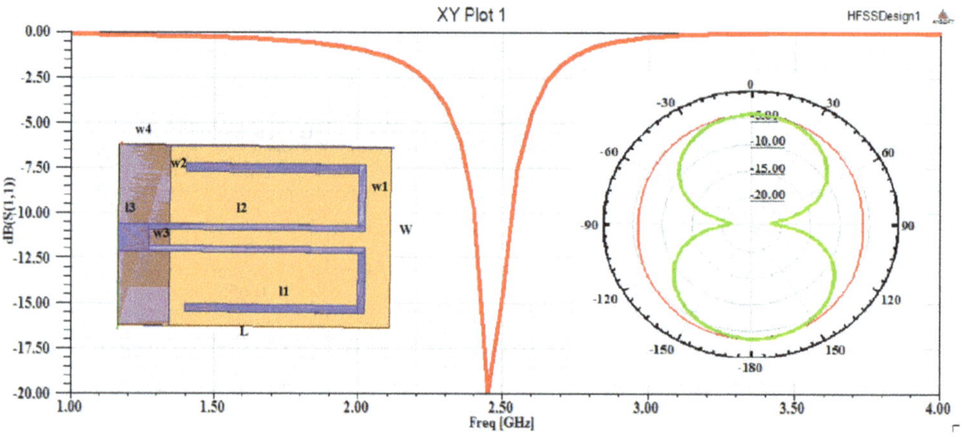


## SAR Calculation (Specific Absorption Rate) for Implanted Antenna

There are two specific absorption rate calculation are there, One is local SAR and another one is average SAR. Local SAR is how much radiation is absorbed by the a particular area of the body and average SAR is average of the local SARs. SAR can be calculated by the equation,

*Figure 17. Simulated Implanted antenna Electric Field Pattern*

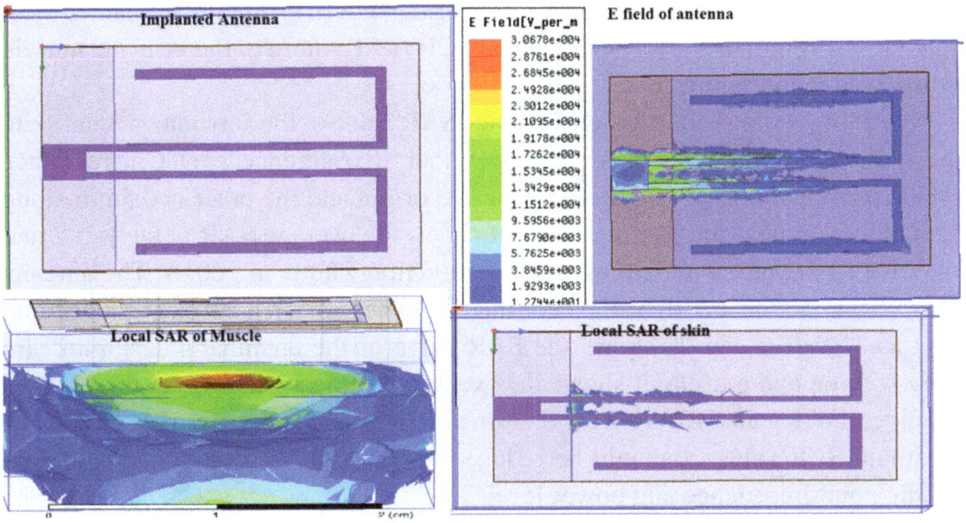


SAR= 1/v(∫σ(r)\E(r)\²/p(r))dr

σ = electrical Conductivity
E = electric field
ρ=density of dielectric material

*Figure 18. Local and Average SAR of antenna implanted on skin*

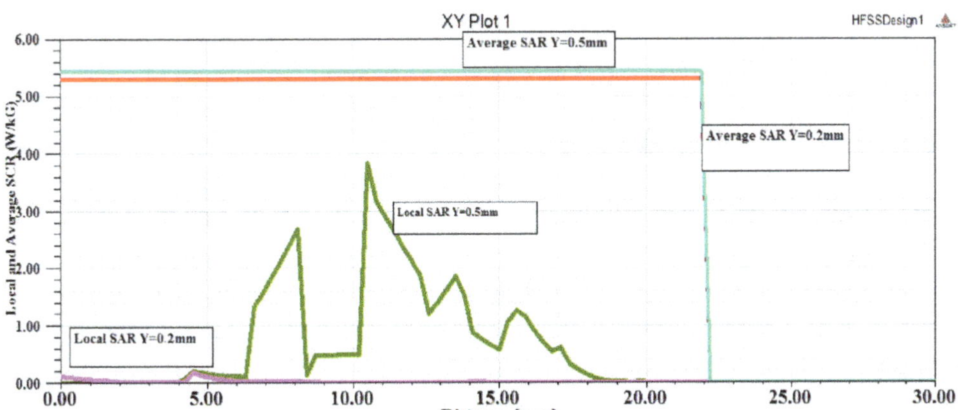

The above implanted antenna SAR value is calculated(Liu & Guo, 2013) under two conditions (i) Antenna implanted with 3mm skin (ii) Antenna implanted with 3mm fat and 20mm muscle. Antenna local SCR field pattern of the skin and muscle is displayed in the figure .17

Figure 18,19 shows the local and average SAR value of the antenna on 3mm skin. The SAR value of the antenna is measured under two reference level. One reference level is 0.2mm along the Y direction near the origin and the other is 0.5mm along the Y direction near the feeding point. It shows the average SAR value is 5.2 and 5.4W/Kg for 0.2mm and 0.5mm along Y direction(Zhu et al., 2016).The antenna gives very good local SAR value near the origin as 0.2W/Kg but near the feeding point as 3.8W/Kg. Fig 20 shows the SAR value of the antenna under 3mm skin along with fat and muscle. It shows the average SAR value is 6 for both the cases and gives 4W/Kg and 40W/Kg for 0.2mm and 0.5mm conditions. This case gives maximum SAR value, it should be reduced to less than 2W/Kg by adjusting the feeding point impedance and power level.

*Figure 19. Local and Average SAR of antenna implanted with skin, fat and muscle*

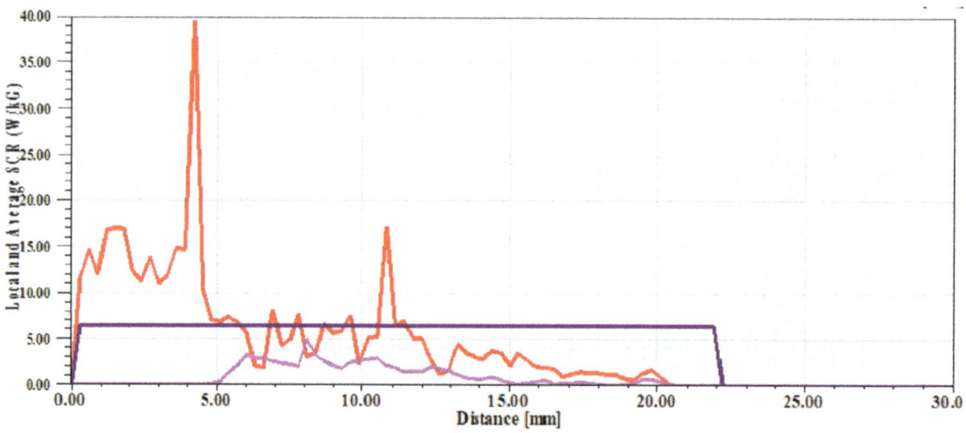


## Optically Controlled Antenna

This type of antenna can be operated by applying a laser beam to the antenna. A photoconductor switch is connected on the antenna for switching purpose. The conductance of the photoconductors with is changed by the applied laser beam to reconfigure the antenna(Xiaomu et al., 2007).This laser beam can be applied from any optical or Li-Fi networks.

*Table 4. Applications of simulated frequency of antenna*

| Sl. No | Switching State | Mode | Mode Frequency [GHz] | S11 [dB] | Applications |
|---|---|---|---|---|---|
| 1. | 00 | Off body | 2.27-2.4 | -32.61 | Body area network and Wi-Fi |
| 2. | 00 | Off body | 4.4407 | -26.74 | 5G spectrum, Mobile applications |
| 3 | 01 | Off body | 3.8305 | -20.74 | Wi -Max, UAV(unmanned Air Vehicle) |
| 4. | 01 | 0ff body | 5.6610 | -27.96 | FCC-ID, Wi-Fi LAN, Radar system |
| 5. | 01 | Off body | 4.6085 | -16.76 | |
| 6 | 10 | Off body | 3.8305 | -25.20 | Wi -Max, UAV(unmanned Air Vehicle) |
| 7. | 10 | Off body | 5.66 | -25.80 | FCC-ID |
| 8 | 11 | Off body | 4.44 | -23.47 | FCC-ID |
| 9. | 00 | Antenna on the body tissue | 3.60 | -15.32 | |
| 10 | 00 | Antenna on the body tissue | 2.7 | -15.90 | |
| 11 | 00 | Antenna on the body tissue | 5 | -15.45 | |
| 12 | 01 | Antenna on the body tissue | 3.7 | -16.8 | |
| 13 | 10 | Antenna on the body tissue | 3.7 | -18.42 | |
| 14 | 11 | Antenna on the body tissue | 3.8 | -17.39 | |
| 15 | 11 | Antenna on the body tissue | 4.4 | -16.90 | |
| 16 | 00 | Implanted antenna(1mm inside) | 2.8 | -16 | |
| 17 | 01 | Implanted antenna(1mm inside) | 3.8 | -15.36 | |
| 18 | 01 | Implanted antenna(1mm inside) | 4.6 | -13.30 | |
| 19 | 10 | Implanted antenna(1mm inside) | 3.1 | -15.35 | |
| 20 | 11 | Implanted antenna(1mm inside) | 4.6 | -16.69 | |
| 21 | 11 | Implanted antenna(1mm inside) | 3.9 | -15.86 | |
| 22 | 11 | Antenna 0.5mm away from body tissue | 3 | -14.07 | |
| 23 | 00 | Antenna 0.5mm away from body tissue | 4.5 | -15.68 | |
| 24 | 01 | Antenna 0.5mm away from body tissue | 3 | -10.9 | |
| 25 | 01 | Antenna 0.5mm away from body tissue | 4.5 | -17.28 | |
| 26 | 10 | Antenna 0.5mm away from body tissue | 3 | -14.7 | |
| 27 | 10 | Antenna 0.5mm away from body tissue | 4.2 | -14.4 | |
| 28 | 11 | Antenna 0.5mm away from body tissue | 3.0337 | -13.63 | |
| 29 | 11 | Antenna 0.5mm away from body tissue | 4.5 | -20.89 | 5G applications |
| 30 | 00 | Antenna h=0.22mm | 5.1 | -13.7 | |
| 31 | 00 | Antenna h=0.22mm | 2.4 | -19.7 | WBAN |
| 32 | 01 | Antenna h=0.22mm | 2.8 | -27.46 | Bio Medical |
| 33 | 10 | Antenna h=0.22mm | 3 | -14 | |
| 34 | 10 | Antenna h=0.22mm | 5.2 | -25 | Wi _ Max |
| 35 | 11 | Antenna h=0.22mm | 2.94 | -36.6 | FCC ID |
| 36 | 00 | Polyester substrate | 5.03 | -10.36 | |
| 37 | 00 | Polyester substrate | 5.5 | -10.32 | |

## Applications

The generated list of frequencies and their applications are listed in the table. The applications are given for only the reflection co-efficient value -20dB.

## FUTURE RESEARCH DIRECTIONS

The antenna size and SAR values can be reduced to wear this antenna along with any medical device. A photoconductive switch can be designed for optically controlled reconfigurable antenna to shift the frequency and pattern.

## CONCLUSION

The inverted and implanted antennas are analyzed and presented here for wireless body area networks and other applications. This antenna can be radiated and reconfigured for minimum two frequency modes in every switching condition. At every switching condition one frequency mode is operating for WBAN and the second mode is operating for other wireless frequency applications. Antenna parameters are simulated and analysed for ON body and OFF body conditions with different position of the human body tissue. The antenna radiation parameters are analyzed for different substrate condition with high gain .A person, who can wear this antenna for receiving or transmitting the signal from body area networks and other wireless applications. This antenna also can be worn easily with cloths.

## ACKNOWLEDGMENT

This research is supported by “Visvesvaraya PhD Scheme for Electronics & IT” Ministry of Electronics and Information Technology, Government of India. Unique awardee number is VISPHD-MEITY-3061.

## REFERENCES

Antoskiewicz, K., & Shafai, L. (2017). Impedance characteristics of circular microstrip patches. *IEEE Transactions on Antennas and Propagation*, *38*(6), 942–946. doi:10.1109/8.55597

Hall, P. S., & Hao, Y. (2006). *Antennas and Propagation for Body-Centric Wireless communications*. Artech House.

Liu, C., Guo, Y. X., & Xiao, S. (2012). Compact dual-band antenna for implantable devices. *IEEE Antennas and Wireless Propagation Letters*, *11*, 1508–1511. doi:10.1109/LAWP.2012.2233705

Liu, C., Guo, Y.-X., & Xiao, S. (2014). Capacitively loaded circularly polarized implantable patch antenna for ISM band biomedical applications. *IEEE Transactions on Antennas and Propagation*, *62*(5), 2407–2417. doi:10.1109/TAP.2014.2307341

Liu, Z. G., & Guo, Y. X. (2005). Compact low-profile dual band metamaterial antenna for body centric communications. *IEEE Antennas and Wireless Propagation Letters*, *14*, 863–866. doi:10.1109/LAWP.2014.2382586

Liu, Z. G., & Guo, Y. X. (2013). Dual band low profile antenna for body centric communications. *IEEE Transactions on Antennas and Propagation*, *61*(4), 2282–2285. doi:10.1109/TAP.2012.2234071

Masood, R., Person, C., & Sauleau, R. (2007). A dual-mode, dual-port pattern diversity antenna for 2.45-GHz WBAN. *IEEE Antennas and Wireless Propagation Letters*, *16*, 1064–1067. doi:10.1109/LAWP.2016.2620724

Shakib & Moghavvemi. (2017). Design of a tri-band off-body antenna for WBAN communication. *IEEE Antennas Wireless Propag. Lett.*, *16*, 210–213.

Simorangir, Yang, Esselle, & Diao. (2017). A varactor tuned frequency reconfigurable fabric antenna embedded in polymer. *IEEE MTT-S International microwave Symposium (IMS),* 204-207.

Simorangkir, Yang, & Esselle. (2018). Method to Realize Robust Flexible Electronically Tunable Antennas Using Polymer Embedded Conductive Fabric. IEEE Transactions on Antennas and Propagation, 66.

Tong, X., Liu, C., Liu, X., Guo, H., & Yang, X. (2018), Switchable ON-/OFF-Body Antenna for 2.45 GHz WBAN Applications. IEEE Transactions on Antennas and Propagation, 66(2).

Xiaomu, H., Yan, S., & Vandenbosch, G. A. E. (2007). Wearable button antenna for dual-band WLAN applications with combined on and offbody radiation patterns. *IEEE Transactions on Antennas and Propagation*, *65*(3), 1384–1387. doi:10.1109/TAP.2017.2653768

Zhu & Guo. (2016). A compact dual-band antenna for wireless body-area network applications. *IEEE Antennas Wireless Propag. Lett.*, *15*, 98–101.

Zhu, X.-Q., Guo, Y.-X., & Wu, W. (2016). A novel dual-band antenna for wireless communication applications. *IEEE Antennas and Wireless Propagation Letters*, *15*, 516–519. doi:10.1109/LAWP.2015.2456039

# Chapter 3
# Survey on Digital Signal Processing for FMCW Radar

**Naseemuddin Ansari**
*Bhagwant University, Ajmer, India*

**Virendra K. Sharma**
*Bhagwant University, Ajmer, India*

**Sanjeev Sharma**
*New Horizon College of Engineering, India*

**Vinod Kumar Singh**
https://orcid.org/0000-0002-0671-0631
*SR Group of Institutions, India*

## ABSTRACT

*In the ongoing years, the radar innovation, once utilized dominatingly in the military, has begun to rise in various regular citizen applications. One of the regions that this innovation showed up is car automation. The danger of crash is incremented as the quantity of vehicles out. Versatile voyage control, pre-crash, and cautioning are a portion of the utilizations of car radar which are as of now being used in numerous autos. The eventual fate of the car business guarantees to offer a completely self-ruling auto that can drive itself with no driver help. These vehicles will require great radar sensors that can give exact data about the encompassing of the vehicle. These sensors will likewise require a figuring stage that can guarantee constant handling of the received signals. It is discovered that digital signal processing comprises of 3D FFT and framework augmentation handling. As an elective preparing stage, a FPGA/ASIC stage can be utilized to actualize of the calculation to meet the ongoing imperatives of car applications.*

DOI: 10.4018/978-1-7998-7611-3.ch003

## I. INTRODUCTION

The Frequency Modulated Continuous Wave (FMCW) radar has numerous preferences over the normally utilized pulse radars and has discovered an assortment of uses. FMCW level radars have been utilized for quite a long time in the regions, for example, route and auto hostile to crash evasion. Attributable to its low transmitting power and high accuracy in range measurement, the FMCW radar is preferably appropriate for level estimation applications and has been utilized for quite a long time. A standout amongst the most interesting highlights of the FMCW level radar is its high precision in range measurement which in order of 1mm. Frequency modulated continuous-wave (FMCW) radars have discovered various applications where the nearness of one target is to be identified. Be that as it may, this kind of radar is more hard to utilize where there are many targets as detection dependent on the analysis of the frequency spectrum of the received signal. No powerful techniques have been created for automatically handling the processing of received signal to identify the targets, particularly if the target return is of an indistinguishable magnitude as clutter level. In prior work, where a FMCW radar was utilized to recognize covered objects the elucidation of the frequency spectrum was performed by visual review of the yield of a spectrum analyzer. Such a technique for translation is for the most part unacceptable. So through this venture we endeavor to discover a strategy for target extraction utilizing advanced digital-signal-processing.

Self-driving cars offer a totally new perspective on the application of the radar technology in the automobiles. Instead of only assisting the driver, the new automotive radars should be capable of taking an active role in the control of the vehicle. As a matter of fact, they will be a key sensor of the autonomous control system of a car.

Radar is preferred over the other alternatives such as sonar or lidar as it is less affected by the weather conditions and can be made very small to decrease the effect of the deployed sensor to the vehicle's aerodynamics and appearance. The Frequency Modulated Continuous Wave (FMCW) radar is a type of radar that offers more points of interest contrasted with the others. It ensures the range and velocity information of the surrounded objects to be detected simultaneously. This information is very crucial for the control system of the self-driving vehicle to provide a safe and collision-free cruise control.

A radar system installed in a car should be able to provide the necessary information to the control system in real-time. It requires to have a base-band processing system which is capable of providing enough computing power to meet the real-time system requirements. The processing system performs digital signal processing on the received signal to extract the useful information such as range and velocity of the surrounded objects.

## II. LITERATURE SURVEY

Axelsson S. R. J. (Axelsson, 2007) This work executes the possibility of an engineered range, where numerous narrowband CW tones are utilized to integrate the impact of a wide band waveform. it talks about the same matched filter, anyway he additionally expounds on another technique appropriate for digital signal processing, wherein an example of the amplitude/phase is taken at each received signal tone, at that point an IDFT is used to convert samples in time-domain. Wehner presumes that MF and IDFT techniques are proportionate. A reflected received signal is correlated with many reference signals. Much work has been devoted to the signal processing of FMCW waveform, Most of them is dealing with the range estimation of the objects. Liu Y. (Liu, Meng, Zhang et al, 2008) Proposes a digital signal processing technique to calculate the range and Doppler dependent on least square error estimation method. This technique only works for linear stepped frequency waveform. It uses a least mean square method to deal with appraise the speeds of targets to compensate range measurement mistake; this examination manages linear frequency steps. Liu Y. (Liu, Meng, Li et al, 2008) Proposes a calculation to gauge the range and Doppler of numerous targets. The calculation requires from the earlier learning of the quantity of focuses for exact estimation. Morelande M. (Morelande et al., 2013) This patent proposes a strong and down to earth calculation to assess the range-Doppler-azimuth of numerous objectives utilizing waveform planning accomplished by substituting between a CW waveform and RSF waveform. Al-Hourani A. (Al-Hourani et al., 2016) Proposes a signal processing strategy to acquire the range and Doppler data by the rehashed transmission of RSF waveforms to recognize moderate swaying emerging from the Doppler impact. Shyr-Long Jeng (Jeng et al., 2014) proposed, Evaluating and ordering vehicle speed are urgent issues in VDs used to assemble traffic information in an ITS. Be that as it may, as noted in Area I, per-vehicle speed estimation by side-looking single-beam microwave detection is by and large off base or unsupported. Likewise, gathering reliable length information from these identifiers is unthinkable as a result of the loud speed gauges given by customary information accumulation to single beam locators. The proposed calculation empowers precise discovery of each target velocity and width by utilizing side-observed one-beam microwave finders. Not at all like ordinary side-looking radar sensors, the proposed strategy utilizes a glance point to acquire more grounded Doppler signals contrasted and customary sensors. Since the proposed conspire couples velocity and distance estimations for the path in which a target is moving, evaluations of target area and development are expressed in a way that makes its completely clear, consistently good in quality and performance. The headings of moving vehicles are likewise decided by up-or down-uprooting Doppler shifts, despite the prevailing circumstances to the driving part of a road. The width of a target can be likewise calculated from its velocity and time duration in the detecting area. In this manner, a solitary radar can proficiently gauge velocity and screen all paths of a movement along a route. Recreations affirmed that the hypothetical range-Doppler FMCW guideline inferred in this examination can be utilized to identify various targets all the while. A financially accessible 15.6-GHz radar and adjusted signal processing programming were utilized for trial to check quality, performance and reliability. The exploratory outcomes demonstrate that the proposed calculation is profoundly precise, especially at low speeds. The estimation error was under 4.5 km/h. Notwithstanding recognizing vehicles, the investigations demonstrated that the present strategy can distinguish feeble moving targets, for example, bicycles and people on foot. In spite of the fact that the proposed calculation displayed satisfactory execution at low speeds, additionally thinks about are expected to build the pertinent target speed extend for the calculation. Diminishing the permanent deviation in the direction of the gaze of one eye edge can decrease outspread velocity yet would likewise debilitate the Doppler speed and speed determination. The arrangement is utilizing a DSP with higher execution capacity to enhance the strength and exactness of the proposed calculation. Jau. Jr. Lin (Lin & Li, 2016) proposed the baseband signal processing algorithm and architecture calculations in the car radar frameworks. To fulfill the short estimation time imperative without expanding the RF front-end stacking, the three-section waveform is received, which fuses the check slope to confirm the reduce data and increment the framework dependability. By given the abstraction details, the same parameters of three-fragment waveform are additionally examined. To conquer the apparition target and covering issue, the observation design and calculations are proposed. The proposed multi-target identification architecture, which is made out of the discovery, the matching, and the confirmation forms, can viably lessen blending time and upgrade the framework unwavering quality. In addition, the proposed multi-target edge estimation process can resolve the covering issue while having low matching time and high unwavering quality. At long last, reproduction and field estimation comes about show that the proposed discovery design and calculations can dependably identify numerous objectives in the car radar frameworks. The proposed design and calculations adjust the execution and the many-sided quality and are to a great degree reasonable to be actualized in a useful car radar framework. Tai Fei (Fei, 2017) proposed a high-determination approach, i.e. the matrix pencil technique (MPM), to the FMCW car radar framework to isolate the neighboring targets, which share comparative parameters, i.e. go, relative speed and azimuth point, and cause covering in the radar range. With a specific end goal to adjust the 1D model of MPM to the 2D territory speed range and all as far as possible the computational cost, some preprocessing steps are proposed to build a novel partition calculation. Younsik Son (Son & Heo, 2018) proposed a novel phantom target evacuation calculation utilizing the between connection between the energetic and downbeat recurrence after the Fast Fourier Transform(FFT). Calculation proficiently recognize the genuine target when the quantity of the apparition target isn't too vast, so our calculation is helpful when joined with other waveform based strategy. Younsik Son (Son & Heo, 2018) proposed a few MTI digital filtering strategies for FMCW radar applications; background suppression, FIR filtering, and IIR filtering. Each of the tried MTI procedures have reasonable applications. background suppression is appropriate for moderate moving targets. In the event that moving target division is required, the FIR and IIR filtering methods are the easiest, one-advance procedures. It demonstrated that the background suppression method without a noteworthy effect on moderate moving targets. Notwithstanding, the procedure isn't adaptable and can't without much of a stretch be utilized for target segmentation dependent on speed. FIR and IIR filters channels are to a great degree adaptable and their plan can be effectively figured out how to actualize FMCW MTI filters and target segment filters. Specifically, IIR filters show efficiency regarding rationale usage, a valuable property for ease frameworks which regularly utilize the FMCW radar engineering.

## III. METHODOLOGY

### A. Radars

Radar which remains for Detection and Ranging, is a system that utilize electromagnetic waves to distinguish and find objects. A common radar system comprises of a transmitter, beneficiary and a digital signal processing (DSP) module. Initially, the transmitter antenna radiates electromagnetic energy in space. If there is an object within the range of the antenna, it will intercept some of the radiated energy and reflect it in multiple directions. Some of the reflected electromagnetic waves will be returned and received by the receiver antenna. After amplification and some signal processing operations, target information such as distance, velocity and direction can be acquired (Sarkar & Palma, 2014).

*Figure 1. Typical signal processing used in FMCW radar*

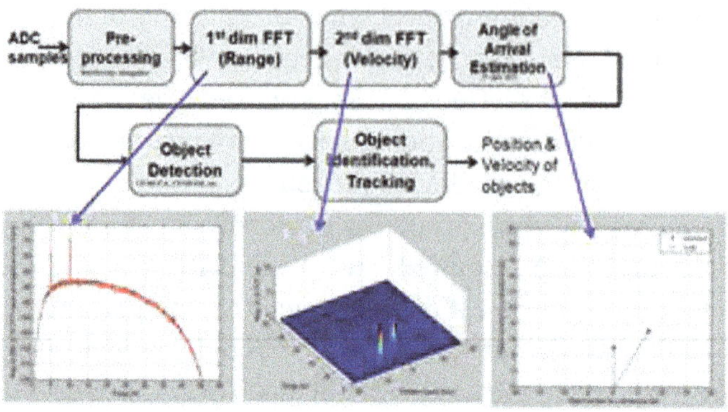


## B. Operation of Radar

The theory of operation of FMCW radar is simple. FMCW radar sends a continuous wave with an increasing frequency. A transmitted wave after being reflected by an object is received by a receiver. Transmitted and received signals are blended to create the signal to be processed by a digital signal processing block. The multiplication process will generate two signals; one with a phase equal to the difference of the multiplied signals, and the other one with a phase equal to the sum of the phases. The sum signal will be filtered out and the difference signal will be processed by the signal processing unit. Figure 2 represents the block diagram of FMCW radar system. Typical digital signal processing flow of FMCW Radar system is shown in Figure 1. The expression for instantaneous FM transmitted chirp in sinusoidal form can be written as

$$S_{tr}(t)= a0_{s}in2\pi[f0t+ Kt2/2^{]} \quad (1)$$

where f0 $_{i}$s the center frequency and K is the chirp rate. The received signal will time delayed version of above equation with various amplitude because of attenuation. Suppose time delay between transmitted and received signal is td, then the equation for the received signal becomes

$$Sr_{(}t)= b0_{s}in2\pi[f0(_{t}\text{-td})+ K(t\text{-td})2_{/}2^{]} \quad (2)$$

The de-ramped signal SIF $_{i}$s the combination of these two signals.

$$SI_{F(}t)= St_{r(}t).\ Sr_{(}t) \quad (3)$$

*Figure 2. FMCW Radar Block Diagram*

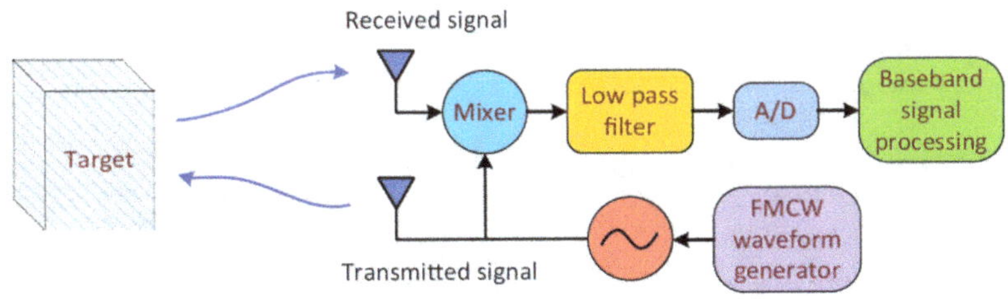


De-ramped signal can be represented as difference and sum of two signal phases. For FMCW just the difference is required in this manner the sum can be overlooked or filtered out. By using Equation (1) and (2) to Equation (3) the de-ramped signal can be represented by Equation (4)

$$S_{IF}(t)= c_0\cos 2\pi[f0_t + Kt.td - Ktd2_{/}^{2}] \quad (4)$$

Here, the second term is the function of time linear phase term corresponding to the instantaneous frequency difference between the transmitted signal and received signal.

From the above equation velocity and range can be respectively expressed as

$$R_0= cT(f_{bu} - f_{bd})/2B \quad (5)$$

$$v_0= c(f_{bu} + f_{bd})/2f_c \quad (6)$$

Above equation shows that bandwidth B is identified with given range resolution and can be composed as

$$B= c/2\Delta R \quad (7)$$

$$T= c/2fc\Delta v \quad (8)$$

*Figure 3. Tx and Rx Waveform of FMCW Radar System*

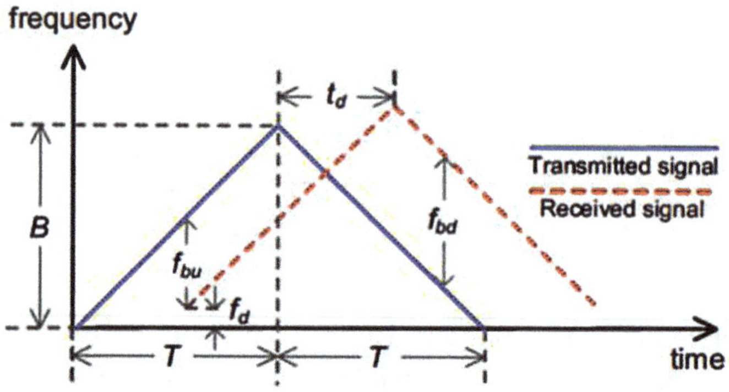


## C. Advantages of FMCW Radar

FMCW radar offers a lot of advantages compared to the other types of radars. These are

- Measure small range with high accuracy.
- Signal processing is performed at relatively low frequency ranges, Capacities well in numerous kinds of climate and air conditions as rain, snow, dampness, haze and dusty condition.
- Little weight and little vitality utilization because of nonappearance of high voltage circuits.

## IV. CONCLUSION

As the number of sensor (Tx/Rx) increases the complexity of the signal processing increases. It is discover after a deliberate search that signal processing comprises of 3D FFT and matrix multiplication. To meet the complexity of the computational requirement of the system. Considering the calculation and the continuous necessity of the application, it was discovered that multi-processor system isn't skilled to meet ongoing imperatives of the application. As an elective handling stage, a FPGA/ASIC stage can be utilized to actualize of the calculation to meet the ongoing imperatives of automotive applications.

## REFERENCES

Al-Hourani, Evans, Farrell, Moran, Martorella, Kandeepan, Skafidas, & Parampalli. (2016). Millimeter-wave CMOS Radar Systems and Techniques. *Academic Press Library in Signal Processing, 7.*

Axelsson, S. R. J. (2007, April). Analysis of Random Step Frequency Radar and Comparison With Experiments. *IEEE Transactions on Geoscience and Remote Sensing*, *45*(4), 890–904. doi:10.1109/TGRS.2006.888865

Fei, T. (2017). A Novel Target Separation Algorithm Applied to The Two- Dimensional Spectrum for FMCW Automotive Radar Systems. *IEEE International Conference on Microwaves, Antennas, Communications and Electronic Systems*. 10.1109/COMCAS.2017.8244791

Jeng, S.-L., Chieng, W.-H., & Lu, H.-P. (2014, April). Estimating Speed Using a Side-Looking Single-Radar Vehicle detector. *IEEE Transactions on Intelligent Transportation Systems*, *15*(2), 607–614. doi:10.1109/TITS.2013.2283528

Lin, & Li. (2016). Design of an FMCW radar baseband signal processing system for automotive application. *SpringerPlus*.

Liu, Y., Meng, H., Li, G., & Wang, X. (2008, August). Range-Velocity Estimation of Multiple Targets in Randomised Stepped-Frequency Radar. *Electronics Letters*, *44*(17), 1032–1034. doi:10.1049/el:20081608

Liu, Y., Meng, H., Zhang, H., & Wang, X. (2008). Motion compensation of moving targets for high range resolution stepped-frequency radar. *Sensors (Basel)*, *8*(5), 3429–3437. doi:10.33908053429 PMID:27879886

Morelande, Mei, & Evans. (2013). *A Method of Target Detection*. cA Patent.

Sarkar, T. K., & Palma, M. S. (2014). A History of the evolution of RADAR. *44th European Microwave Conference*, 734-737. 10.1109/EuMC.2014.6986539

Son, Y., & Heo, S. W. (2018). A novel multitarget detection algorithm for automotive FMCW radar. *2018 International Conference on Electronics, Information, and Communication (ICEIC)*, 1-3.

# Chapter 4
# A Diamond Cut Leather Substrate Antenna for BAN (Body Area Network) Application

**Raghav Agrawal**
*University of Technology, India*

**Pramod Sharma**
*University of Technology, India*

**Anurag Saxena**
*SR Group of Institutions, India*

## ABSTRACT

*In this chapter, a diamond cut shape antenna has been simulated under CST software using twisting material (i.e., leather). The dielectric constant of leather material is 1.655. For antenna to be twist, leather substrate is used. The return loss of this newly design diamond cut shaped antenna is -35 dB at 5.5467 GHz resonant frequency. This type of antenna can be useful in BAN (body area network) applications. In this chapter, simulated results like return loss graph, 3-D pattern, and polar plot are presented.*

DOI: 10.4018/978-1-7998-7611-3.ch004

## INTRODUCTION

The simulation of leather antenna can be done in three steps. Firstly, we have to make a ground of copper material having dimensions on x-axis and y-axis. Secondly, after the ground, substrate is imposed on it having some dimensions and lastly patch is designed on leather substrate with copper material (Lemey et al., 2014; Saxena, 2019; Saxena & Khare, 2019; Saxena et al., 2019; Van Torre et al., 2010).

In the town, after a complicated multiple reflection or scattering effect, the polarization of the propagating radio wave may change significantly. These are used in high performance aircrafts, radar, missiles and other spacecraft (Lui et al., 2013; Saxena, Singh, Mohini et al, 2018; Saxena & Singh, 2018a; Saxena & Singh, 2018b; Singh et al., 2018; Singh & Saxena, 2018).

It has many advantages such as its light weight, simple structure, ease of addition and less cost. Microstrip antenna requires very less space for installation as these are simple and small in size. The only space these need is the gap for feed line which is placed at the back of the ground plane (Agilent, 2005; Roundy, 2003; Saxena, Singh, Kumar et al, 2018; Saxena & Singh, 2019a; Saxena & Singh, 2019b).

There are a variety of methods for enhancing the bandwidth of microstrip antenna by expands the material depth, utilizing low dielectric material, using numerous feeding techniques and impedance matching (Shukla et al., 2015).

Both the bandwidth and the thickness of the antenna is contradictory assets i.e. enhancement in bandwidth increases the size & thickness of presented antenna (Hall & Hao, 2006).

## DESIGN CONFIGURATION OF PRESENTED ANTENNA

The design of Diamond cut leather antenna and ground is shown in figure 1. The dimension ground made by copper is 60 x 60 mm having 0.0038 mm height. A patch square is designed on leather substrate with different diamond cut. The height of leather antenna is 2 mm. Table 1 shows the various parameters of presented antenna.

## RESULTS AND DISCUSSION

After simulating the diamond cut shape leather antenna various results are generated and shown in figure given below. Figure 2 shows the return loss Vs frequency graph in which 5.5467 GHz resonant frequency is generated. Figure 3 shows the radiation pattern in 3-D format of presented antenna in which around 6.382 directivity is generated on 5.567 GHz resonant frequency. Also Figure 4 shows the radiation pattern in polar plot of presented antenna in which main lobe magnitude is 5.9 dBi, main lobe direction is 34.0 deg, angular width (3dB) is 44.9 deg and side lobe level is -13.7 db.

*Figure 1. Front view and back view of presented antenna*

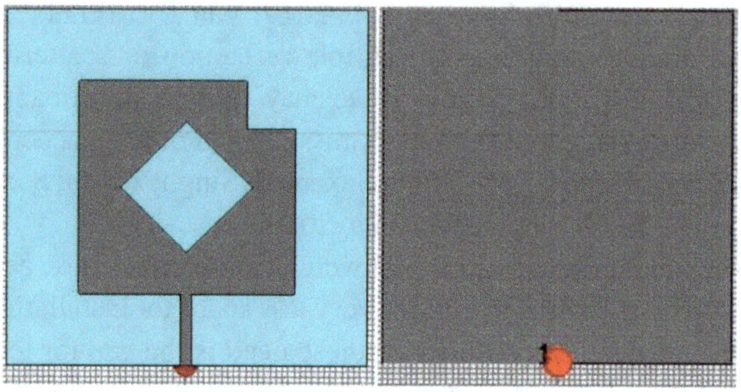

*Table 1. Parameters of Presented Antenna*

| Parameters | Value |
|---|---|
| Relative permittivity ($\varepsilon_r$) | 1.655 |
| Ground Dimension | 60 * 60 mm |
| Resonant Frequency | 5.5467 GHz |
| Substrate Dimension | 60 * 60 mm |
| Strip Line Dimension | 1 * 30 mm |
| Height of Leather Material | 2 mm |

*Figure 2. Return loss Vs frequency simulated graph*

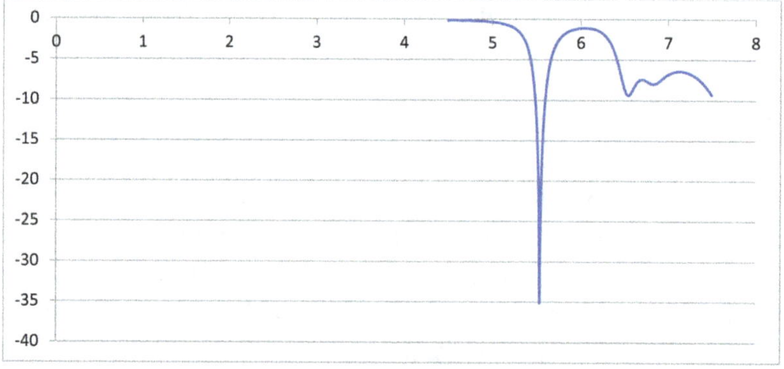

*Figure 3. Radiation Pattern in 3-D format of presented antenna*

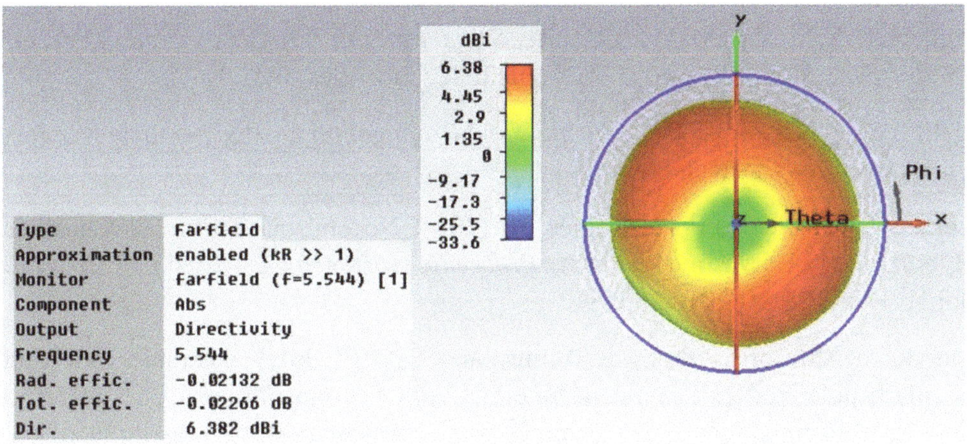


*Figure 4. Radiation pattern in polar plot of presented antenna*

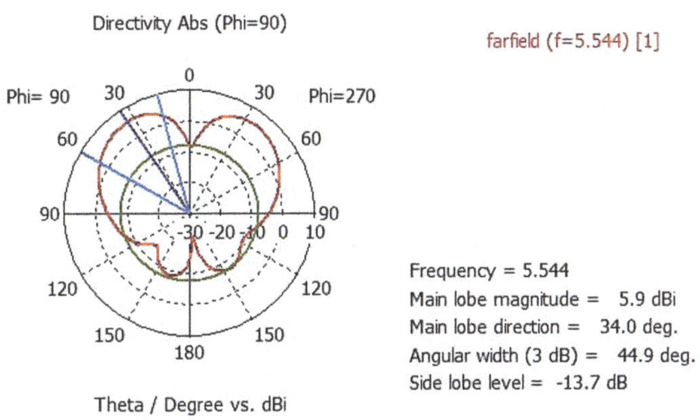


## CONCLUSION

For BAN Network, a twisting material antenna has been designed and synthesis on a leather substrate with relative permittivity of 1.655. The suggest antenna has a return loss of -35 dB. The overall performance of presented antenna is better than the performance of square shape type antenna with FR-4 material. Further this antenna is used for BAN applications and can also be used for energy harvesting.

## REFERENCES

Agilent, H. S. M. S. (2005). *-285x Series Surface Mount Zero Bias Schottky Detector Diodes-Data Sheet 5898-4022EN*. Agilent Technologies, Inc.

Hall, P. S., & Hao, Y. (2006). *Antennas and Propagation for Body-centric Wireless Communications*. Artech House.

Lemey, S., Declercq, F., & Rogier, H. (2014, November). Textile Antennas as Hybrid Energy-Harvesting Platforms. *Proceedings of the IEEE*, *102*(11), 1833–1857. doi:10.1109/JPROC.2014.2355872

Lui, K. W., Murphy, O. H., & Toumazou, C. (2013, July). wearable wideband circularly polarized textile antenna for effective power transmission on a wirelessly-powered sensor platform. *IEEE Transactions on Antennas and Propagation*, *61*(7), 3873–3876. doi:10.1109/TAP.2013.2255094

Roundy, S. J. (2003). *Energy scavenging for wireless sensor nodes with a focus on vibration to electricity conversion* (PhD Thesis). University of California, Berkeley, CA.

Saxena. (2019). A La Shape Antenna for High Frequencies Application. In *Design and Optimization of Sensors and Antennas for Wearable Devices: Emerging Research and Opportunities*. IGI Global.

Saxena & Khare. (2019). A Hammer type Textile Antenna with partial circle ground for wide band applications. In *Design and Optimization of Sensors and Antennas for Wearable Devices: Emerging Research and Opportunities*. IGI Global.

Saxena, Raizada, Gautam, & Khare. (2019). Efficient Rectenna Circuit for Wireless Power Transmission. In *Design and Optimization of Sensors and Antennas for Wearable Devices: Emerging Research and Opportunities*. IGI Global.

Saxena, Singh, Kumar, & Ansari. (2018). A Concentric Double hole Shape Microstrip Patch Antenna at 3.12 GHz Resonant Frequency. In *Advanced Computation and Telecommunication (ICACAT)*. IEEE.

Saxena, Singh, Mohini, Bhardwaj, Chae, Sharma, & Bhoi. (2018). Rectenna Circuit at 6.13 GHz to operate the sensors devices. *International Journal of Engineering and Technology, 7*(2.33), 644-646.

Saxena, A., & Singh, V. K. (2018a). A Moon-Strip Line Antenna for Multi-Band Applications at 5.44 GHz Resonant Frequency. *4th International Conference on Advances in Electrical, Electronics, Information, Communication and Bio-Informatics (AEEICB-18)*.

Saxena, A., & Singh, V. K. (2018b). A Watch E-Cut Shape Textile Antenna for WB Applications. *Journal of Microwave & Technologies*, *5*(1), 29–32.

Saxena, A., & Singh, V. K. (2019a). Antenna for S-Band Application Utilizing Flexible Material. *International Journal of Broadband Cellular Communication.*, *5*(1), 1–4p.

Saxena, A., & Singh, V. K. (2019b). Triangle Cut Circle Shape Bending Antenna for Wide Band Applications. *Journal of Microwave Engineering and Technology.*, *6*(2), 14–17p.

Shukla, S. S., Verma, R. K., & Gohir, G. S. (2015). Investigation of the effect of Substrate material on the performance of Microstrip antenna. *4th International Conference on Reliability, Infocom Technologies and Optimization*, 1-3. 10.1109/ICRITO.2015.7359350

Singh, Saxena, Khare, Shayka, Chae, Sharma, & Bhoi. (2018). Power Harvesting through flexible rectenna at dual resonant frequency for low power devices. *International Journal of Engineering and Technology*, *7*(2.33), 647-649.

Singh, V. K., & Saxena, A. (2018). Two Parabolic Shape Microstrip Patch Antenna for Single Band Application. *Journal of Microwave & Technologies*, *5*(1), 33–36.

Van Torre, P., Vallozzi, L., Rogier, H., & Verhaevert, J. (2010). Diversity textile antenna systems for firefighters. *Antennas and Propagation (EuCAP), Proceedings of the Fourth European Conference.*

# Chapter 5
# Analysis of Bandwidth of Microstrip Antenna for Broad Band Applications

**Deepak Niranjan**
*Dr. A. P. J. Abdul Kalam Technical University, India*

**Satyendra Swarnkar**
*SR Group of Institutions, India*

## ABSTRACT

*The anticipated antenna is studied and presented. The anticipated antenna is connected to probe feed when designed on IE3D software. The bandwidth, gain, VSWR, and reflection coefficient have been found which is of 77.71% bandwidth, and gain is of 4.1 dBi. This antenna is compact, lightweight, and suitable for portable devices. By using some advanced techniques, the antenna provides dual bandwidth and triple bandwidth. The anticipated antenna provided large bandwidth for and is utilized for broadband applications.*

## 1. INTRODUCTION

The recent development of antenna technology is key feature for wireless communication area and the patch antennas execute all the needs. Microstrip antennas are very simple in designing as these are low profile. The least parameters are needed to design these antennas. The double sided printed circuit board is utilized to fabricate such antennas. To get large bandwidth, the substrate of low permittivity is appropriate whereas the substrate of higher permittivity gives smaller size of

DOI: 10.4018/978-1-7998-7611-3.ch005

antenna (Mourad & Essaaidi, 2014; Naresh & Singh, 2020; Rachana et al., 2018; Sharma et al., 2018; Singh Nikhil et al., 2018; Xu & Li, 2012).

The better efficiency is received when the height of substrate increases, however at the same time surface waves are generated which is unwanted. The enhancement of bandwidth and efficiency are the major objective which has been studied in anticipated research article (Abhishek et al., 2018; Grilo & Correra, 2015; Jain et al., 2013; Kunal et al., 2018; Manju & Singh, 2018; Niharika et al., 2018; Poornima et al., 2018; Princi et al., 2018; Rahul et al., 2018; Rawat & Sharma, 2014). Conventional microstrip antennas have few demerits such as narrow bandwidth and low efficiency. By using some advance techniques, the antenna provides dual bandwidth and triple bandwidth. But nowadays dual band and triple band is not sufficient parameters for designing the microstrip antennas (). There are many techniques which is not only utilized for the enhancement of bandwidth but also cost effective.

## 2. ANTENNA DESIGN

The top view of proposed antenna on IE3D is presented in Figure 1. The presented antenna was made of ground with the dimension of 46mm× 56mm and patch dimensions are 36.4mm×46.4mm. In this design, glass epoxy substrate has been utilized having the 1.6mm having and permittivity ($\varepsilon_r$) of 4.2. Three dimensional structure of anticipated antenna is shown in Figure 2.

## 3. OUTCOMES OF ANTENNA

The simulation of anticipated antenna is carried out IE3D software and the results such as back loss, impedance band width, gain and efficiency are obtained. The return loss vs. frequency graph is represented in figure 3 which gives 77.71% bandwidth. The VSWR is represented in figure 4 and Gain is shown in figure 5.

### 3.1 Antenna Gain

Since the gain and bandwidth product of the antenna is constant. As the gain increases bandwidth reduce and converse is also holds true. The gain vs. frequency plot of the suggested microstrip antenna is represents in the figure 5. The gain of proposed microstrip antenna is 4.1 dBi.

*Figure 1. Top view of proposed antenna on IE3D*

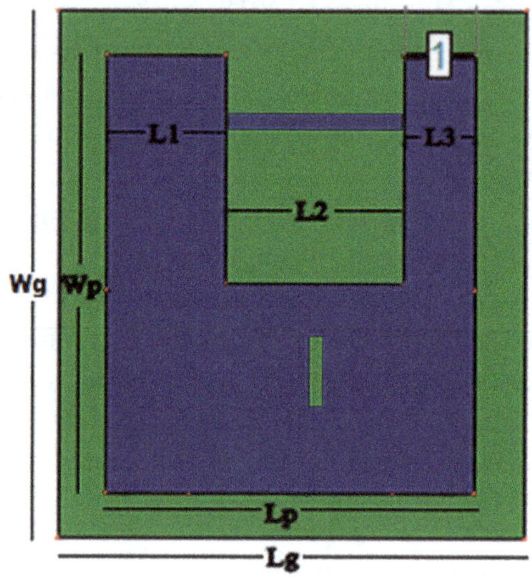


## 3.2 Directivity

The plot of directivity vs. frequency for suggested microstrip antenna is shown in figure 5.7. The maximum directivity of presented antenna 5.2 dBi.

*Figure 2. 3-D View of proposed microstrip antenna*

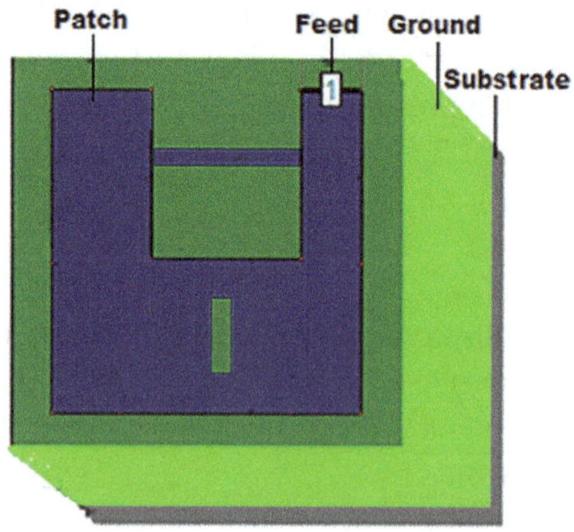

*Table 1. Parameters of proposed antenna*

| Parameter for proposed antenna | Value (mm) |
|---|---|
| Width of ground plane ($W_g$) | 56 |
| Length of ground plane ($L_g$) | 46 |
| Width of patch ($W_p$) | 46.4 |
| Length of patch ($L_P$) | 36.4 |
| Substrate height (h) | 1.6 |
| Length of left arm of patch ($L_1$) | 11.7 |

*Figure 3. Return Loss vs. Frequency plot of antenna*

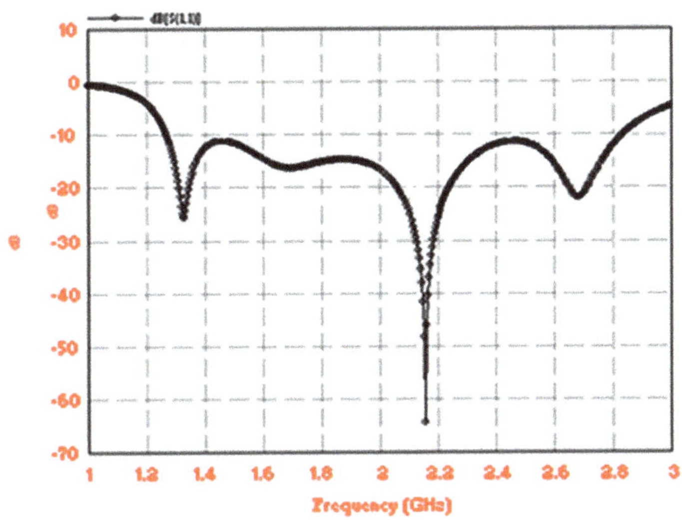


## 4. EXPERIMENTAL RESULTS

The front view and back view of fabricated antenna is shown in figure 7. Figure 8 shows the comparative study of simulated and measured results of proposed antenna. It is found that there is slight difference between simulated and experimental results due to some limitations such as fabrication method and losses..

*Figure 4. VSWR plot for U -shaped microstrip patch antenna*

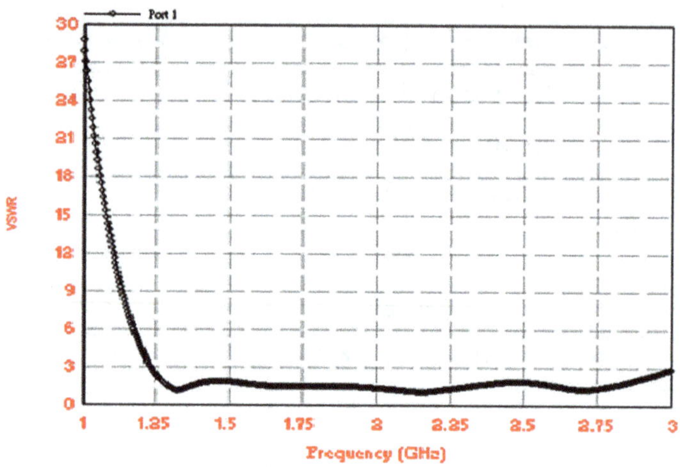


*Figure 5. Gain Vs. Frequency plot of suggested antenna*

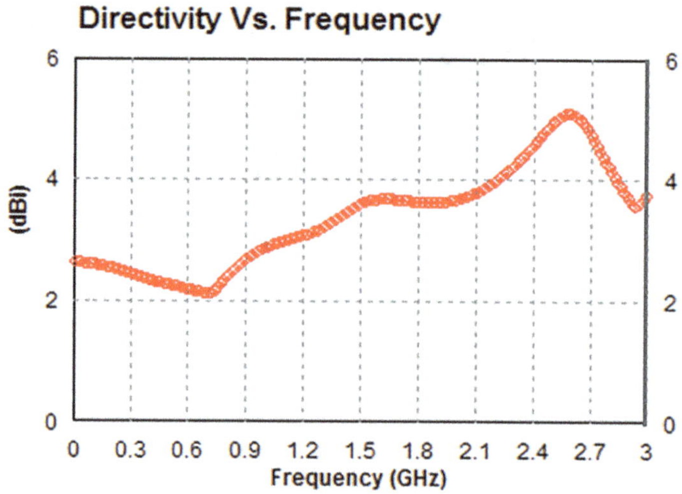


## 5. CONCLUSION

The proposed antenna design gives the bandwidth of 77.71%, gain of 4.1 dBi, and directivity is 5.2 dBi. The proposed antenna is simple, compact size and low cost and id suitable for broad band application.

*Figure 6. Simulated Directivity of proposed antenna*

*Figure 7. Fabricated design of suggested patch antenna on PCB: (a) Front view of fabricated antenna (b) Back view of fabricated antenna*

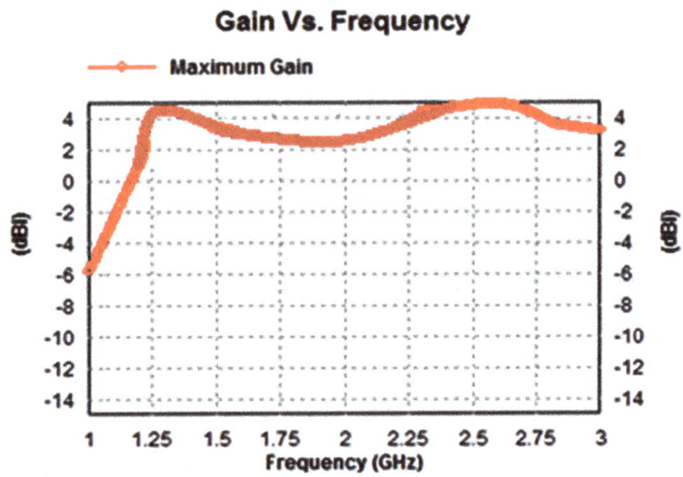

*Figure 8. Comparative study of simulated and measured results of proposed antenna*

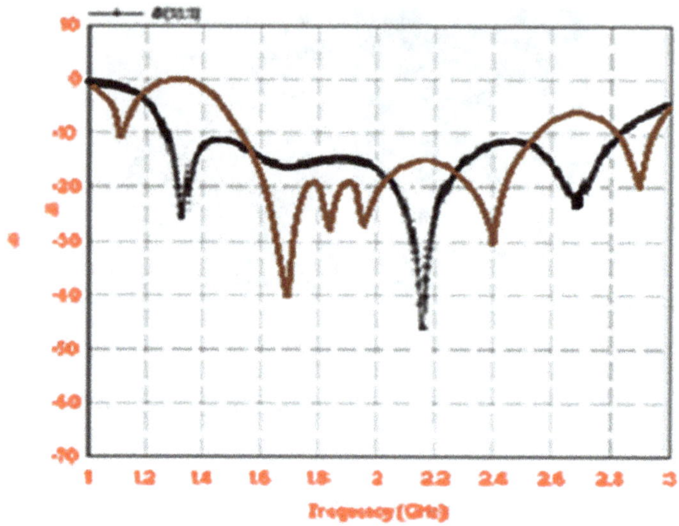


## REFERENCES

Abhishek, S., Sanjeev, G., Singh, V. K., & Bhoi, A. K. (2018). Design of Permanent Magnet Synchronous Generator for Wind Energy Conversion System. Lecture Notes in Electrical Engineering, 435, 23-32.

Ashish, K., Archana, L., Singh, V. K., & Bhoi, A. K. (2018). Estimation of Frequency Band of Microstrip Antenna (MSA) with Radial Basis Function (RBF). *Lecture Notes in Electrical Engineering*, *462*, 385–391. doi:10.1007/978-981-10-7901-6_42

Deepak & Pathak. (2013). A study on inverted T shaped micro strip antenna at different frequencies. *International Journal of Engineering and Computer Science*, *2*(11), 3180-3183.

Grilo, M., & Correra, F. S. (2015, July). Rectangular patch antenna on textile substrate fed by proximity coupling. *Journal of Microwaves, Optoelectronics and Electromagnetic Applications*, *14*, 103–112.

Hu, B., Gao, G. P., He, L. L., Cong, X. D., & Zhao, J. N. (2016). Bending and on-arm effects on a wearable antenna for 2.45 GHz body area network. *Antennas Wireless Propag IEEE Letter*, *15*, 378–381. doi:10.1109/LAWP.2015.2446512

Jain, S., Singh, V. K., & Ayub, S. (2013). Band Width and Gain Optimization of a Wide Band Gap Coupled Patch Antenna. *International Journal of Engineering Sciences & Research Technology*, 649-652.

Jiang, Z. H., Brocker, D. E., Sieber, P. E., & Werner, D. H. (2014). A compact, low-profile metasurface-enabled antenna for wearable medical body-area network devices. *IEEE Transactions on Antennas and Propagation*, *62*(8), 402–4030. doi:10.1109/TAP.2014.2327650

Kunal, L., Archana, L., & Singh, V. K. (2018). Wide Band Triangular Patch Textile Antenna with Partial Ground Plane. *Advances in Intelligent Systems and Computing*, *632*, 577–583. doi:10.1007/978-981-10-5520-1_52

Loni, Singh, Tripathi, Chae, Sharma, & Bhoi. (2018). Improvement of the performance of wearable textile antenna. *IACSIT International Journal of Engineering and Technology*.

Manju, D., & Singh, V. K. (2018). Antenna for Wireless Area Network and Bluetooth Application. Lecture Notes in Electrical Engineering, 462, 377-384.

Mourad, M., & Essaaidi, M. (2014). A dual ultra wide band slotted antenna for C and X-band application. *Progress in Electromagnetics Research Letters*, *47*, 91–96. doi:10.2528/PIERL14052103

Naresh, B., & Singh, V. K. (2020). Integration of RF Rectenna with thin film solar cell to power wearable electronics. *International Journal of Microwave and Wireless Technologies*. doi:10.100740009-020-00877-4

Niharika, S., Singh, V. K., Rahul, S., & Prakash, S. J. (2018). Microstrip Textile Antenna with Jeans Substrate with Applications in S-Band. Lecture Notes in Electrical Engineering, 462, 369-376.

Pathak, R. S., Singh, V. K., & Ayub, S. (2012). Dual band Microstrip Antenna for GPS/ WLAN/WiMax Applications. *International Journal of Emerging Trends in Engineering and Development, 7*(2), 154-159.

Poornima, S., Singh, V. K., Archana, L., & Bhoi, A. K. (2018). Design and Analysis of Microstrip Antenna Using Multilayer Feed-Forward Back-Propagation Neural Network (MLPFFBP-ANN). *Lecture Notes in Electrical Engineering*, *462*, 393–398. doi:10.1007/978-981-10-7901-6_43

Princi, T., Upendra, T., Singh, V. K., & Bhoi, A. K. (2018). Solution of Economic Load Dispatch Problems through Moth Flame Optimization Algorithm. *Lecture Notes in Electrical Engineering*, *462*, 287–294. doi:10.1007/978-981-10-7901-6_31

Rachana, K., Singh, V. K., Singh, N. K., & Anurag, S. (2018). A Compact Pentagonal Textile Microstrip Antenna for Wide Band Application. Lecture Notes in Electrical Engineering, 443, 793-799.

Rahul, Singh, Singh, Prakash, & Bhoi. (2018). Multi Resonant Textile Antenna with Partial Ground for Multiband Applications. Lecture Notes in Electrical Engineering, 462, 359-367.

Rawat, S., & Sharma, K. K. (2014). A compact broadband microstrip patch antenna with defected ground structure for C-band applications. *Central European Journal of Engineering*, *4*(3), 287–292.

Saxena, Singh, Lohia, & Faisal. (2020). A Jawbreaker-Shaped Fragmentary Ground Textile Flexible Antenna. *International Journal of Electronics, Communications, and Measurement Engineering*.

Saxena, A., & Singh, V. K. (2018). A Moon-Strip Line Antenna for Multi-Band Applications at 5.44 GHz Resonant Frequency. *Fourth International Conference on Advances in Electrical, Electronics, Information, Communication and Bio-Informatics (AEEICB)*. 10.1109/AEEICB.2018.8480998

Sharma, Ashish, & Vinod. (2018). Design of Circularly Polarized Antenna with Different Iterations for UWB Applications. Lecture Notes in Electrical Engineering, 443, 441-447.

Singh, Ali, & Ayub. (2010). Design of compact triple band microstrip antenna for wireless communication. International Journal of Electronics & Communication Engineering, 3(1), 323-330.

Singh, Ali, Singh, & Ayub. (2013). Dual Band Microstrip Antenna for UMTS/WLAN/WIMAX Applications. *International Conference on Communication Systems and Network Technology (CSNT-2013)*, 47 – 50.

Singh, N., Singh, A. K., & Singh, V. K. (2015). Design & Performance of Wearable Ultra Wide Band Textile Antenna for Medical Applications. *Microwave and Optical Technology Letters, 57*(7), 1553-1557. doi:10.1002/mop.29131

Singh, V. K., Dhupkariya, S., & Bangari, N. (2017). Wearable Ultra Wide Dual Band Flexible Textile Antenna for WiMax/WLAN Application. *International Journal of Wireless Personal Communications, 95*(2), 1075–1086. doi:10.100711277-016-3814-7

Singh, V. K., & Singh, N. K. (2018). Rectenna design for electromagnetic energy harvesting and wireless power transfer. *IACSIT International Journal of Engineering and Technology*.

Singh Nikhil, Niraj, Zakir, Singh, & Bhoi. (2018). Inset Fed Circular Microstrip Antenna with Defected Ground. Lecture Notes in Electrical Engineering, 443, 605-611.

Xu & Li. (2012). A Dual band Microstrip Antenna for Wearable Application. *IEEE (ISAPE) Conference Proceeding*, 109-112.

Yadav & Singh. (2019). Design of U-shape with DGS circularly polarized wearable antenna on fabric substrate for WLAN and C-Band applications. *Journal of Computational Electronics, 18*(3), 1103–1109. . doi:10.100710825-019-01342-2

Yadav, A., Kumar Singh, V., Kumar Bhoi, A., Marques, G., Garcia-Zapirain, B., & de la Torre Díez, I. (2020). Wireless Body Area Networks: UWB Wearable Textile Antenna for Telemedicine and Mobile Health Systems. *Micromachines, 11*(6), E558. doi:10.3390/mi11060558

Yu, X. J., Tang, Q., Liang, M., & Xin, H. (2014). Study of power efficiency of non-foster impedance matching for electrically small antenna. *IEEE Conf Publ.* 10.1109/USNC-URSI.2014.6955512

# Chapter 6
# Bandwidth Analysis of Dual-Feed Slotted Antenna Using Artificial Neural Networks

**Archana Lala**
*SR Group of Institutions, India*

**Kunal Lala**
*Raj Kumar Goel Institute of Technology, India*

**Vinod Kumar Singh**
https://orcid.org/0000-0002-0671-0631
*SR Group of Institutions, India*

## ABSTRACT

*In this chapter, artificial neural network is used for the estimation of bandwidth of a dual feed microstrip antenna. The MLPFFBP-ANN and RBF-ANN are used to implement the neural network model. The simulated values for training and testing the neural network are obtained by simulating the antenna on IE3D software. The results obtained by using ANNs and IE3D simulation are compared and are found quite acceptable, and also it is concluded that RBF network is more accurate and fast as compared to back propagation algorithm of MLPFFBP. The anticipated is applicable to operate in triple band from 2.208GHz-5.35GHz, 2.358GHz-2.736GHz, and 3.815GHz-5.143GHz. The antenna is also fabricated with FR-4 glass epoxy material. The experimental results, simulated results of IE3D, and simulated results of neural network are compared.*

DOI: 10.4018/978-1-7998-7611-3.ch006

# 1. INTRODUCTION

Neuro models are computationally substantially more effective then EM models once they are prepared with dependable taking in information get from a fine model by either EM reproduction or estimation. The Neuro models can be utilized for proficient precise advancement and planned inside the scope of preparing (Gao et al., 2017; Hassoun, 1999; Haykin, 2003; Kumar & Ray, 2003; Zeland Software Inc, n.d.). The Method of Moments (MOM) based IE3D programming has been utilized to create preparing and test information for the ANN. It is a computational EM test system dependent on Method of Moments numerical techniques. It is investigation that a 3D and multi layer structure of general shapes feed point must be situated at point on the fix where the information impedance of fix coordinated the feed for the particular full recurrence. The arrival misfortune is recorded and that feed point is chosen as the ideal one where the RL is most negative for example not exactly - 10db. It is anything but difficult to demonstrate and simple to coordinate by controlling the test feed arranges (Jain et al., 2019; Rashmi et al., 2018; Singh Nikhil, Singh Vinod, Saxena et al, 2018; Thakre & Singhal, 2009; Wu et al., 2015).

Microstrip reception apparatus' are utilized in remote correspondence because of its light loads low profile, ease, mandate with high transmission productivity and simplicity of incorporation with microwave circuit and convenient correspondence equipments, that is the reason it discovers place in much applications since 1970's. Anyway slotted cross antenna as the disadvantage of restricted transfer speed, limited power, weak polarization, however analysts have been proposed and researched numerous procedures to overcome the downsides, by using thick substrates, adjusted ground plane and utilization of many feed methods and impedance coordinating strategies and the utilization of various resonators(Jamshed, 2015; Mourad & Essaaidi, 2014; Naresh & Singh, 2020; Poornima, 2018; Rachana et al., 2018; Sharma et al., 2018; Singh Nikhil, Sharma Niraj, Zakir et al, 2018; Xu & Li, 2012).

The proposed antenna give triple bandwidth of 83.10%, 28.12% and 14.84% covering the frequency range of. 2.208GHz - 5.35GHz, 2.358GHz - 2.736GHz and 3.815GHz - 5.143GHz which is best suitable for WLAN (Hyper line & W Line) UWB (low band Europe & Japan) and Wimax,Worldwide Interoperability for micro access IEEE 802.11g and Long range surveillance for Radar application system

# 2. DESIGNING AND DATA CREATION

Figure 1 describes the geometry of the conventional antenna. A radio wire has 33.4 mm×40.6 mm altered ground plane and 23.8 mm×31 mm of slotted cross patch measurements. The microstrip receiver has the length L the width W, which is shown in Figure 1.

*Figure 1. Geometry of proposed microstrip antenna*

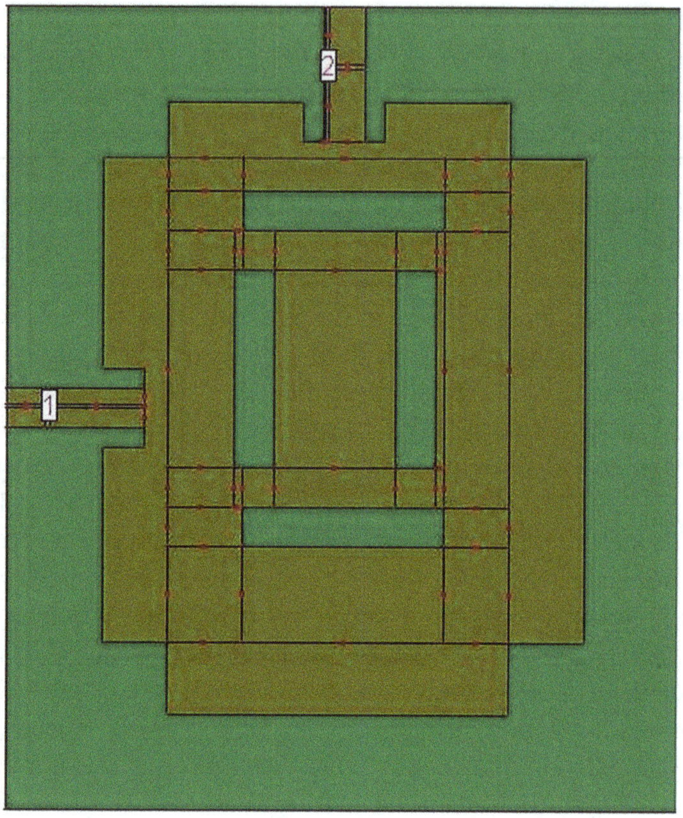


The substrate chosen for this plan is FR-4 glass epoxy with dielectric consistent ($\varepsilon_r$) = 4.4 and tallness of the substrate (h) = 1.6 mm. The fringing field is increased by using low dielectric constant thus the radiated power is increased. The feed is given at the point (x, y). The anticipated antenna is simulated on IE3D software to get the sufficient date for ANN. Figure 1 shows the format of a coaxial test feed slotted antenna (Abhishek et al., 2018; Grilo & Correra, 2015; Kunal et al., 2018; Manju & Singh, 2018; Niharika et al., 2018; Princi et al., 2018; Rahul et al., 2018; Rawat & Sharma, 2014). By differing the feed position of Port-1 and Port-II of the anticipated antenna the training date have been generated.

## 2.1 ANN Model

The ANN model has been produced for opened cross patch receiver as appeared in Figure 2. The feed forward system has been used to ascertain the data transfer capacity of the antenna by placing, substrate dielectric constant ($\varepsilon_r$) and substrate width (h). This is characterized as examination ANN model. So by giving the different probes at the input of ANN model, the large bandwidth is received without complex calculation.

*Figure 2. Artificial Neural Network Model*

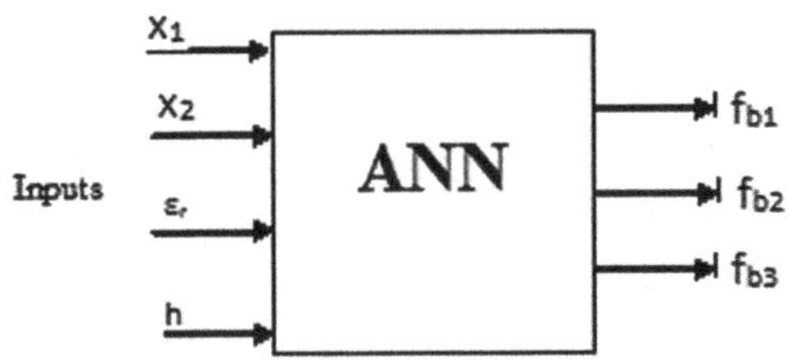


## 3. MULTILAYER PERCEPTRON FEED FORWARD BACK PROPOGATION NEURAL NETWORK

In the network there are 2 input neurons within the input layer, and 25 hidden neuron within the hidden layer and three output neuron within the output layer as shown in Figure 3. The input to the network is x coordinate of probe position of patch antenna and therefore the output of the network is bandwidth (Ashish et al., 2018; Jain et al., 2013; Pathak et al., 2012; Singh et al., 2010; Singh et al., 2013). The training time is 9 seconds and training performs in 200 iterations. So as to judge the performance of the proposed MLPFFBP ANN based model for the structure of patch antenna simulation results are obtain using IE3D simulator and generated 61 input and output training patterns and 10 input and output test patterns to validate the model. During the training process the neural network automatically adjusts its weights and threshold values so the error between predicted and sampled outputs is minimized (Naresh & Singh, 2020; Rachana et al., 2018). The transfer function preferred is tansig and purelin within the architecture. Figure 5 shows the training performance of the developed neural model for proposed antenna using LM training algorithm. Model is trained in 200 epochs and training time was 7 seconds. In table 1, the resonant frequencies obtained using IE3D Software and using L-M Algorithm for various test patterns are compared and therefore the Mean Square Error has been calculated. From the table it is evident that the result obtained from IE3D and

ANN tool is very closed by and hence given accurate result after several trainings the length and width of the patch is kept constant and the probe position is being changed and the network is trained for the same adjustment. Further it is seen in table 3 the measured result by the network analyzer have almost the same frequency band and bandwidth as obtained from the simulator. The ANN tool is just used to study the bandwidth of microstrip antenna which is in good agreement with the results obtained from Zealand IE3D software. It has been observed that in MLPFFBP model, 200 epochs are utilized to reduce MSE level up to a low value 1.00e-$^{07}$for three layers MLPFFBP with Levenberg-Marquardt (LM) training algorithm using Tansig as a transfer function that is shown in Figure 4.

*Figure 3. Three Layer MLFFBP Network Architecture*

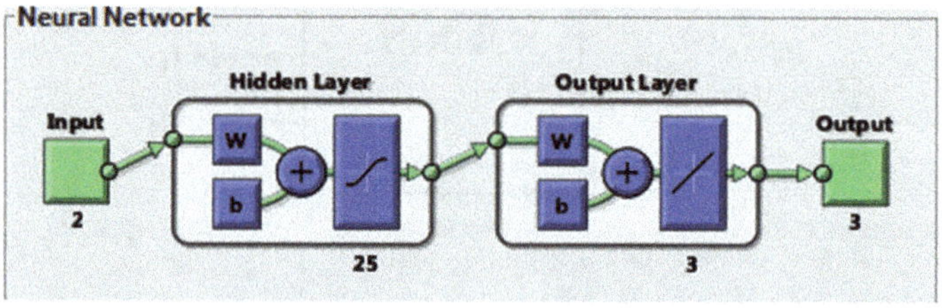


*Table 1. Comparison of Results obtained using IE3D and MLPFFBP using Levenberg-Marquardt algorithm*

| Probe 1 (x, y) | Probe 2 (x, y) | FB1 IE3D GHz | FB2 IE3D GHz | FB3 IE3D GHz | BW1 IE 3D GHz | BW2 IE 3D GHz | BW3 IE 3D GHz | FB1 MLPFFBP GHz | FB2 MLPFFBP GHz | FB3 MLPFFBP GHz | BW1 MLPFFBP GHz | BW2 MLPFFBP GHz | BW3 MLPFFBP GHz |
|---|---|---|---|---|---|---|---|---|---|---|---|---|---|
| 2.0, 20.375 | 15.9, 37.6 | 3.139 | 0.378 | 1.268 | 83.1 | 14.84 | 28.12 | 3.112 | 0.375 | 1.291 | 82.39 | 14.72 | 28.63 |
| 2.2, 20.375 | 16.1, 37.6 | 2.705 | 0.369 | 1.27 | 67.07 | 14.48 | 28 | 2.689 | 0.37 | 1.249 | 66.678 | 14.48 | 27.54 |
| 2.4, 20.375 | 16.3, 37.6 | 2.385 | 0.36 | 1.278 | 56.59 | 14.12 | 28.06 | 2.382 | 0.36 | 1.277 | 56.512 | 14.12 | 28.04 |
| 2.6, 20.375 | 16.5, 37.6 | 2.216 | 0.355 | 1.293 | 51.43 | 13.90 | 28.33 | 2.229 | 0.353 | 1.295 | 51.732 | 13.80 | 28.36 |
| 2.7, 20.375 | 16.6, 37.6 | 2.161 | 0.354 | 1.301 | 49.70 | 13.88 | 28.49 | 2.162 | 0.355 | 1.303 | 49.723 | 13.91 | 28.53 |
| 2.8, 20.375 | 16.7, 37.6 | 2.096 | 0.351 | 1.308 | 47.89 | 13.77 | 28.64 | 2.096 | 0.350 | 1.308 | 47.883 | 13.73 | 28.64 |
| 2.9, 20.375 | 16.8, 37.6 | 2.034 | 0.352 | 1.337 | 46.19 | 13.80 | 29.29 | 2.014 | 0.354 | 1.334 | 45.736 | 13.88 | 29.22 |
| 3.0, 20.375 | 16.9, 37.6 | 1.964 | 0.360 | 1.320 | 44.41 | 13.96 | 28.94 | 1.956 | 0.358 | 1.327 | 44.229 | 13.86 | 29.08 |
| 3.1, 20.375 | 17.0, 37.6 | 1.895 | 0.357 | 1.349 | 42.68 | 14.00 | 29.57 | 1.895 | 0.359 | 1.348 | 42.673 | 14.06 | 29.54 |
| 3.2, 20.375 | 17.1, 37.6 | 1.822 | 0.356 | 1.363 | 40.87 | 14.11 | 29.91 | 1.821 | 0.355 | 1.327 | 40.836 | 14.05 | 29.12 |

*Figure 4. No. of epochs to achieve minimum MSE in case of MLPFFBP*

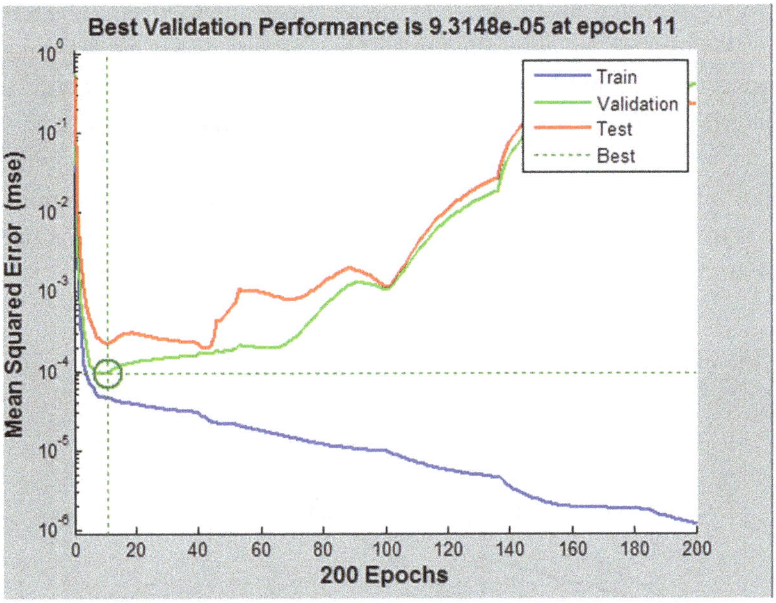


## 4. RADIAL BASIS FUNCTION NETWORK

In this paper, radial basis function (RBF) neural networks is employed to investigate the parameters of patch antenna. Radial Basis Function Network may be a feed forward neural network with one single hidden layer that used radial basis activation function for hidden neurons. RBF network are applied for various microwave modeling purposes. The RBF neural network has both are supervised and unsupervised components to its learning. It consists of three layers of neurons-input, hidden and output. The hidden layer neuron represents a series of centers within the input file space. Each of those centers has activation functions typically Gaussian. The activation depends on the gap between the presented input vector and also the centre. The generation of centers and their widths is done using an unsupervised k-Means clustering algorithm. The centers and also the widths created by these algorithms then form the burden of device of the hidden layer which remains unchanged once the clustering has been done. A typical RBF network is given in Figure 5.

In the RBF network the spread value is chosen as 0.1 which provides the simplest accuracy the network was trained with 61 samples and tested with 10 samples. Within the structure there's one input and one output was used for the analysis ANN as shown in Figure 6, RBF networks are more fast and effective as compared to MLPFFBP for proposed antenna design. The RBF network automatically adjust the quantity of processing elements within the hidden layer till the define accuracies reached. The training algorithm is unsupervised k-mean clustering theorem. The RBF neural network is also utilized to study the microstrip antenna and it has an input layer, a hidden layer and an output layer. The neuron hidden in the layer contains Gaussian transfer function whose outputs are inversely proportional to the distance from the center of the neuron. It is established from table 2 that RBF is given result not only accurate but fast also. The presented RBF network has performed training in less epochs then in MLPFFBP so it is concluded that RBF architecture is better from MLPFFBP and its accuracy is up to 99.97%, 99.80.% and 99.81%

*Figure 5. Radial Basis Function Neural Network approach*

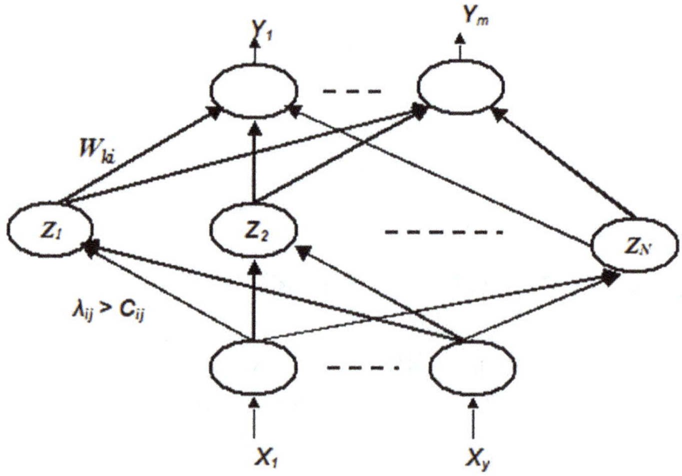


*Figure 6. Radial Basis Function Network*

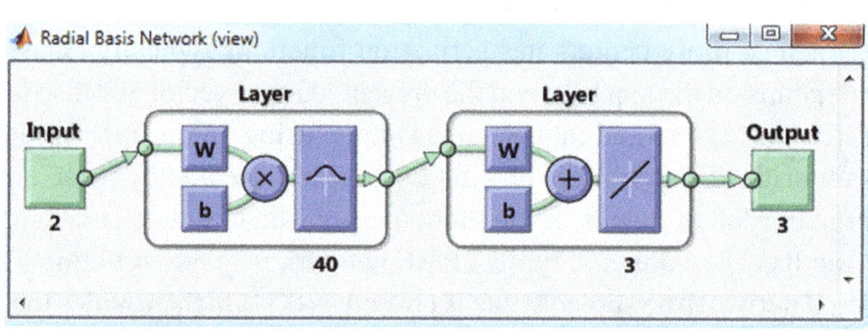


Figure 7 shows the training performance of the developed neural model using RBF network. Model is trained in 40 epochs. It's clear that RBF network is far much faster than feed forward network since RBF network is trained in fewer epochs then feed forward network (200 epochs using LM algorithm). In table 2 bandwidth obtained using IE3D software and using RBF network for various test pattern are compared and therefore the mean square error has been calculated which is sort of zero for all the test pattern.

*Figure 7. No. of epochs to achieved min Mean Square Error level in a case of RBF network*

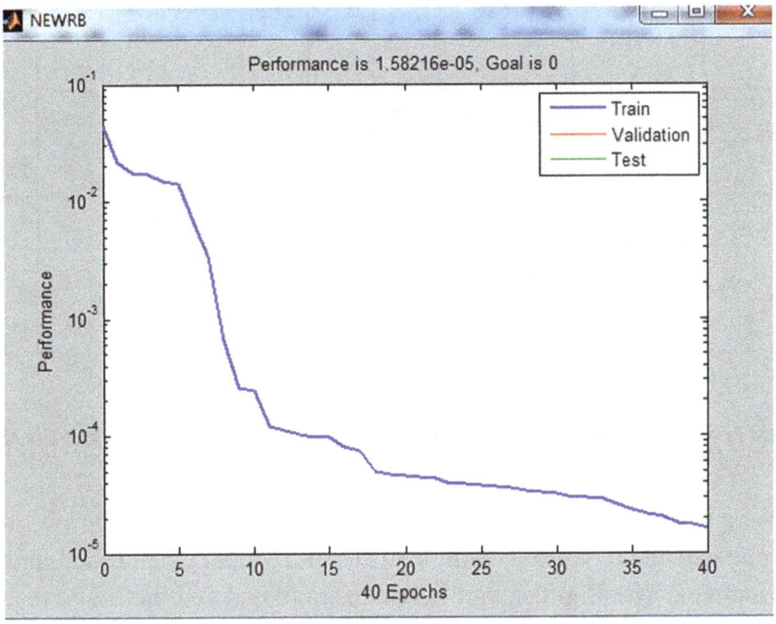


## 5. RESULT AND DISCUSSION

The return loss plot is shown in Figure 8 and the results are depicted in table 2 which is about -35.83db for port-I and -17.37db & -14.67 db for port-II. The proposed antenna resonates at 2.95 GHz frequency and has frequency range from 2.208GHz to 5.35GHz, 2.358GHz to 2.736GHz and 3.815GHz to 5.143GHz therefore it is suitable for WLAN, UWB (low band Europe & Japan) and WiMax application (Deepak & Pathak, 2013; Hu et al., 2016; Jiang et al., 2014; Loni et al., 2018; Singh et al., 2015; Singh et al., 2017; Yadav et al., 2020; Yadav et al., 2019; Yu et al., 2014).

*Table 2. Result obtain using IE3D and RBF ANN algorithm for the analysis of bandwidth*

| Probe 1 (x, y) | Probe 2 (x, y) | FB1 IE3D GHz | FB2 IE3D GHz | FB3 IE3D GHz | BW1 IE 3D GHz | BW2 IE3D GHz | BW3 IE3D GHz | FB1 RBF GHz | FB2 RBF GHz | FB3 RBF GHz | BW1 RBF GHz | BW2 RBF GHz | BW3 RBF GHz |
|---|---|---|---|---|---|---|---|---|---|---|---|---|---|
| 2.0,20.375 | 15.9, 37.6 | 3.139 | 0.378 | 1.268 | 83.1 | 14.84 | 28.12 | 3.139 | 0.378 | 1.268 | 83.10 | 14.84 | 28.12 |
| 2.2,20.375 | 16.1, 37.6 | 2.705 | 0.369 | 1.27 | 67.07 | 14.48 | 28 | 2.691 | 0.369 | 1.2563 | 66.72 | 14.48 | 27.70 |
| 2.4,20.375 | 16.3, 37.6 | 2.385 | 0.36 | 1.278 | 56.59 | 14.12 | 28.06 | 2.381 | 0.359 | 1.2775 | 56.50 | 14.08 | 28.05 |
| 2.6, 20.375 | 16.5, 37.6 | 2.216 | 0.355 | 1.293 | 51.43 | 13.90 | 28.33 | 2.216 | 0.355 | 1.294 | 51.43 | 13.9 | 28.35 |
| 2.7, 20.375 | 16.6, 37.6 | 2.161 | 0.354 | 1.301 | 49.70 | 13.88 | 28.49 | 2.161 | 0.354 | 1.301 | 49.71 | 13.88 | 28.48 |
| 2.8, 20.375 | 16.7, 37.6 | 2.096 | 0.351 | 1.308 | 47.89 | 13.77 | 28.64 | 2.095 | 0.351 | 1.308 | 47.87 | 13.77 | 28.64 |
| 2.9, 20.375 | 16.8, 37.6 | 2.034 | 0.352 | 1.337 | 46.19 | 13.80 | 29.29 | 2.023 | 0.352 | 1.336 | 45.95 | 13.8 | 29.28 |
| 3.0, 20.375 | 16.9, 37.6 | 1.964 | 0.360 | 1.320 | 44.41 | 13.96 | 28.94 | 1.962 | 0.359 | 1.322 | 44.36 | 13.92 | 28.98 |
| 3.1, 20.375 | 17.0, 37.6 | 1.895 | 0.357 | 1.349 | 42.68 | 14.00 | 29.57 | 1.890 | 0.357 | 1.350 | 42.57 | 14 | 29.59 |
| 3.2, 20.375 | 17.1, 37.6 | 1.822 | 0.356 | 1.363 | 40.87 | 14.11 | 29.91 | 1.822 | 0.356 | 1.363 | 40.87 | 14.11 | 29.91 |

## 6. FABRICATED ANTENNA AND EXPERIMENTAL RESULTS

The photograph of hardware design is shown in Figure 9. Figure 10 shows a comparison of both simulated and measured result in terms of frequency band and bandwidth of proposed antenna. The slight change between simulated and measured result is due to fabrication tolerance, humidity, and temperature effect. However, simulated and measured shows a good correlation (Saxena et al., 2020; Saxena & Singh, 2018). The optimization through neural network model shows a good correlation with measured and simulated result of proposed antenna. It is clear from the graph that accuracy achieved using RBF Networks is much better than the accuracy achieved using Levenberg-Marquardt training Algorithm of MLPFFBP Networks.

*Figure 8. Return loss ($S_{11}$) of anticipated antenna*

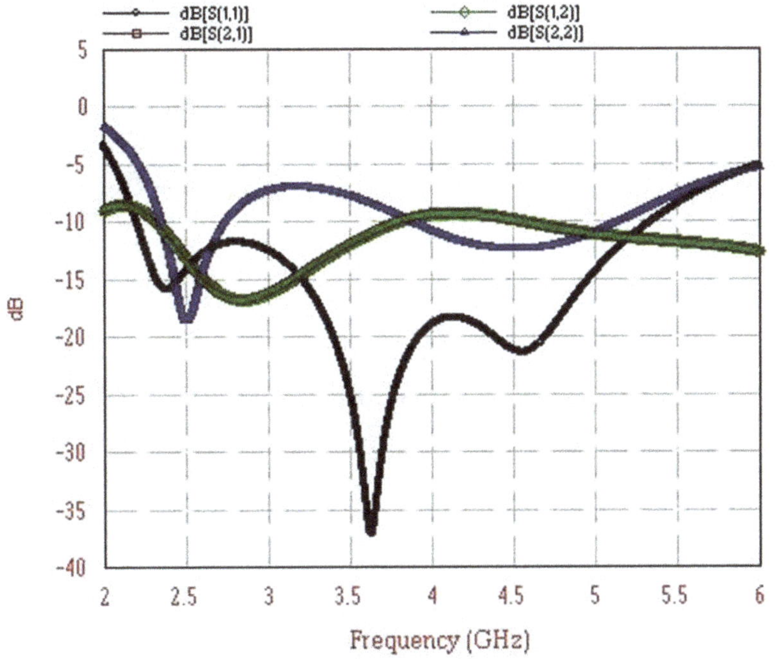


*Figure 9. Prototype of anticipated antenna*

*Figure 10. Simulated and Measured Results*

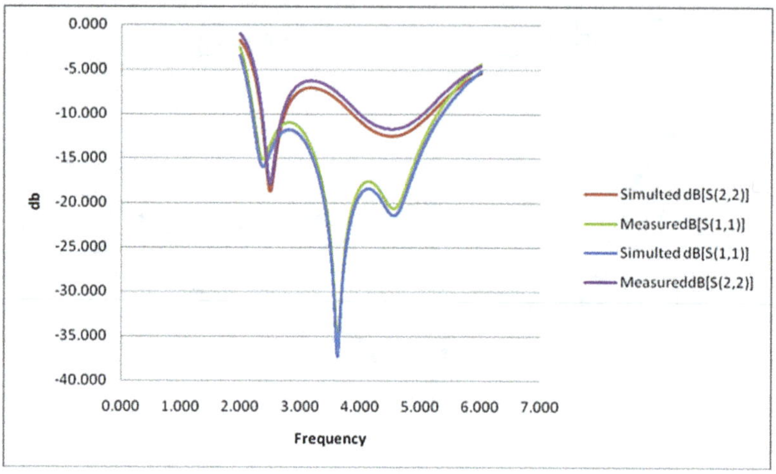


*Table 3. Measured and Simulated Results*

| | Measured Result | Simulated Result (IE3D) | Simulated Result (ANN) | |
|---|---|---|---|---|
| | | | MMLPFFBP | RBF |
| Frequency Band 1 | 3.041 | 3.139 | 3.112 | 3.139 |
| Bandwidth 1 | 81.15% | 83.10% | 82.39% | 83.10% |
| Frequency Band 2 | 0.331 | 0.378 | 0.375 | 0. 378 |
| Bandwidth 2 | 13.04% | 14.80% | 14.72% | 14.80% |
| Frequency Band 3 | 1.037 | 1.268 | 1.291 | 1.268 |
| Bandwidth 3 | 23% | 28.12% | 28.63% | 28.12% |

## 7. CONCLUSION

This paper presents a design of Dual Inset feed slotted cross patch microstrip antenna with line probe dual feed for WLAN and WiMax application. FR-4 Glass epoxy substrate with dielectric constant 4.4 is used which gives a wide bandwidth of 83.10%, 28.12% and 14.80%. In this work ANN is used as a tool to study the frequency band and bandwidth of Microstrip Antenna. The results obtain from IE3D and those obtain from ANN are in good agreement and shows almost 99.6% with MLPFFBP and 99.99% accuracy with RBF. The training and test set has been designed with the data obtain from IE3D simulator.

## REFERENCES

Abhishek, S., Sanjeev, G., Singh, V. K., & Bhoi, A. K. (2018). Design of Permanent Magnet Synchronous Generator for Wind Energy Conversion System. Lecture Notes in Electrical Engineering, 435, 23-32.

Ashish, K., Archana, L., Singh, V. K., & Bhoi, A. K. (2018). Estimation of Frequency Band of Microstrip Antenna (MSA) with Radial Basis Function (RBF). *Lecture Notes in Electrical Engineering*, *462*, 385–391. doi:10.1007/978-981-10-7901-6_42

Deepak & Pathak. (2013). A study on inverted T shaped micro strip antenna at different frequencies. *International Journal of Engineering and Computer Science*, *2*(11), 3180-3183.

Gao, P., Hu, B., Tian, X.-L., Zhao, Q.-L., & Zhang, B.-T. (2017, April). Experimental study of a wearable aperture-coupled patch antenna for wireless body area network. *Microwave and Optical Technology Letters*, *59*(4), 761–766. doi:10.1002/mop.30408

Grilo, M., & Correra, F. S. (2015, July). Rectangular patch antenna on textile substrate fed by proximity coupling. *Journal of Microwaves, Optoelectronics and Electromagnetic Applications*, *14*, 103–112.

Hassoun, M. H. (1999). *Fundamentals of Artificial Neural Network*. Prentice Hall of India.

Haykin, S. (2003). *Neural Networks* (2nd ed.). PHI.

Hu, B., Gao, G. P., He, L. L., Cong, X. D., & Zhao, J. N. (2016). Bending andon-arm effects on a wearable antenna for 2.45 GHz body area network. *Antennas Wireless Propag IEEE Lett*, *15*, 378–381. doi:10.1109/LAWP.2015.2446512

Jain, A. K., Kanaujia, B. K., Dwari, S., Pandey, G. P., & Singh, D. K. (2019, June). Wideband circularly polarized magnetoelectric dipole antenna with I-slot for C-band applications. *Journal of Computational Electronics*, *18*(2), 660–670. doi:10.100710825-018-01295-y

Jain, S., Singh, V. K., & Ayub, S. (2013). Band Width and Gain Optimization of a Wide Band Gap Coupled Patch Antenna. *International Journal of Engineering Sciences & Research Technology*, 649-652.

Jamshed, A. (2015). A Novel Wide Band Microstrip-Line-Fed Antenna with Defected Ground for CP Operation. *Progress In Electromagnetics Research C*, *58*, 169–181. doi:10.2528/PIERC15052305

Jiang, Z. H., Brocker, D. E., Sieber, P. E., & Werner, D. H. (2014). A compact, low-profile metasurface-enabled antenna for wearable medicalbody-area network devices. *IEEE Transactions on Antennas and Propagation*, *62*(8), 402–4030. doi:10.1109/TAP.2014.2327650

Kumar, & Ray. (2003). *Broadband Microstrip antennas*. Artech House.

Kunal, L., Archana, L., & Singh, V. K. (2018). Wide Band Triangular Patch Textile Antenna with Partial Ground Plane. *Advances in Intelligent Systems and Computing*, *632*, 577–583. doi:10.1007/978-981-10-5520-1_52

Loni, Singh, Tripathi, Chae, Sharma, & Bhoi. (2018). Improvement of the performance of wearable textile antenna. *IACSIT International Journal of Engineering and Technology*.

Manju, D., & Singh, V. K. (2018). Antenna for Wireless Area Network and Bluetooth Application. Lecture Notes in Electrical Engineering, 462, 377-384.

Mourad, M., & Essaaidi, M. (2014). A dual ultra wide band slotted antenna for C and X-band application. *Progress in Electromagnetics Research Letters*, *47*, 91–96. doi:10.2528/PIERL14052103

Naresh, B., & Singh, V. K. (2020). Integration of RF Rectenna with thin film solar cell to power wearable electronics. *International Journal of Microwave and Wireless Technologies*.

Niharika, S., Singh, V. K., Rahul, S., & Prakash, S. J. (2018). Microstrip Textile Antenna with Jeans Substrate with Applications in S-Band. Lecture Notes in Electrical Engineering, 462, 369-376.

Pathak, R. S., Singh, V. K., & Ayub, S. (2012). Dual band Microstrip Antenna for GPS/ WLAN/WiMax Applications. *International Journal of Emerging Trends in Engineering and Development, 7*(2), 154-159.

Poornima, S. (2018). *Design and Analysis of Microstrip Antenna Using Multilayer Feed-Forward Back-Propagation Neural Network (MLPFFBP-ANN). Lecture Notes in Electrical Engineering* , 462.

Princi, T., Upendra, T., Singh, V. K., & Bhoi, A. K. (2018). Solution of Economic Load Dispatch Problems through Moth Flame Optimization Algorithm. *Lecture Notes in Electrical Engineering*, *462*, 287–294. doi:10.1007/978-981-10-7901-6_31

Rachana, K., Singh, V. K., Singh, N. K., & Anurag, S. (2018). A Compact Pentagonal Textile Microstrip Antenna for Wide Band Application. Lecture Notes in Electrical Engineering, 443, 793-799.

Rahul, Singh, Singh, Prakash, & Bhoi. (2018). Multi Resonant Textile Antenna with Partial Ground for Multiband Applications. Lecture Notes in Electrical Engineering, 462, 359-367.

Rashmi, S., Singh, V. K., & Puneet, K. (2018). A Compact CPW-Fed Defected Ground Microstrip Antenna for Ku Band Application. *Lecture Notes in Electrical Engineering*, *443*, 231–237. doi:10.1007/978-981-10-4765-7_24

Rawat, S., & Sharma, K. K. (2014). A compact broadband microstrip patch antenna with defected ground structure for C-band applications. *Central European Journal of Engineering*, *4*(3), 287–292.

Saxena, Singh, Lohia, & Faisal. (2020). A Jawbreaker-Shaped Fragmentary Ground Textile Flexible Antenna. *International Journal of Electronics, Communications, and Measurement Engineering*.

Saxena, A., & Singh, V. K. (2018). A Moon-Strip Line Antenna for Multi-Band Applications at 5.44 GHz Resonant Frequency. *2018 Fourth International Conference on Advances in Electrical, Electronics, Information, Communication and Bio-Informatics*. 10.1109/AEEICB.2018.8480998

Sharma, Yadav, & Singh. (2018). Design of Circularly Polarized Antenna with Different Iterations for UWB Applications. Lecture Notes in Electrical Engineering, 443, 441-447.

Singh, Ali, & Ayub. (2010). Design of compact triple band microstrip antenna for wireless communication. *International Journal of Electronics & Communication Engineering, 3*(1), 323-330.

Singh, Ali, Singh, & Ayub. (2013). Dual Band Microstrip Antenna for UMTS/WLAN/WIMAX Applications. *International Conference on Communication Systems and Network Technology (CSNT-2013)*, 47 – 50.

Singh, Singh, & Singh. (2015). Design & Performance of Wearable Ultra Wide Band Textile Antenna for Medical Applications. *Microwave and Optical Technology Letters, 57*(7), 1553-1557.

Singh, V. K., Dhupkariya, S., & Bangari, N. (2017). Wearable Ultra Wide Dual Band Flexible Textile Antenna for WiMax/WLAN Application. *International Journal of Wireless Personal Communications, Springer*, *95*(2), 1075–1086. doi:10.100711277-016-3814-7

Singh Nikhil, Sharma Niraj, Zakir, Singh, & Bhoi. (2018). Inset Fed Circular Microstrip Antenna with Defected Ground. Lecture Notes in Electrical Engineering, 443, 605-611.

Singh Nikhil, Singh Vinod, Saxena, Bhoi, Garg, & Sherpa. (2018). A Compact Slotted Textile Patch Antenna for Ultra-wide Band Application. Lecture Notes in Electrical Engineering, 443, 53-59.

Thakre, V. V., & Singhal, P. K. (2009). Band width analysis by introducing slots in microstrip antenna design using ANN. *Progress in Electromagnetics Research M. Pier M*, *9*, 107–122. doi:10.2528/PIERM09093002

Wu, C.-M., Syu, J.-W., & Liu, W.-C. (2015). Dual-band slotted patch antenna with defective ground forWLAN/WiMAX applications. *Progress in Electromagnetics Research Letters*, *53*, 1–6. doi:10.2528/PIERL15022503

Xu & Li. (2012). A Dual band Microstrip Antenna for Wearable Application. *IEEE (ISAPE) Conference Proceeding*, 109-112.

Yadav, A., Kumar Singh, V., Kumar Bhoi, A., Marques, G., Garcia-Zapirain, B., & de la Torre Díez, I. (2020). Wireless Body Area Networks: UWB Wearable Textile Antenna for Telemedicine and Mobile Health Systems. *Micromachines, 11*(6), E558. . doi:10.3390/mi11060558

Yadav, A., Singh, V. K., & Mohan, H. (2019, September). Design of U-shape with DGS circularly polarized wearable antenna on fabric substrate for WLAN and C-Band applications. *Journal of Computational Electronics*, *18*(3), 1103–1109. doi:10.100710825-019-01342-2

Yu, X. J., Tang, Q., Liang, M., & Xin, H. (2014). Study of power efficiency of non-foster impedance matching for electrically small antenna. *IEEE Conf Publ*, 130. 10.1109/USNC-URSI.2014.6955512

Zeland Software Inc. (n.d.). *IE3D: MoM-based EM simulator*. http:// www.zeland.com

# Chapter 7
# Design and Analysis of Wearable Antennas

**Mahesh Kumar Aghwariya**
*THDC Institute of Hydropower Engineering and Technology, New Tehri, India*

**Amit Kumar**
*THDC Institute of Hydropower Engineering and Technology, New Tehri, India*

**Ragini Sharma**
*KIET Group of Institutes, Ghaziabad, India*

## ABSTRACT

*This chapter presents the various designing methods and challenges associated with wearable antennas, selecting the designing material, various fabrication techniques, and implementation methods on the fabric. Wearable antennas have gained popularity in recent years due to their unmatched properties and unique features. The wearable antenna is capable of providing effective wireless communication, tracking, and sensing. These types of antenna need to be conformal when used on numerous parts of the human body, and they need to be lightweight, flexible, and must be implemented easily on the flexible material. These antennas must be capable of operating with maximum efficiency on the human body. These requirements make the wearable antenna design challenging. They also need to be compact in size, better in coupling with the body, and must be capable of handling issues associated with the device when the target is moving.*

DOI: 10.4018/978-1-7998-7611-3.ch007

## INTRODUCTION OF ANTENNA

An antenna is a device that is used to send and receive the electromagnetic signal. It is an electrical device that takes electric signals and converts while transmitting the signals and covert radio waves into the electrical signals while receiving the signals. While transmitting the electromagnetic wave transmitter supply electric current to the antenna terminal at some specified frequency and the antenna terminal radiates the received signals into space in the form of em waves. At the receiver end where antenna works as a receiver, the antenna captures a small power of the em wave and generates a small voltage between the terminals, this voltage further given to the receiver for amplification.(Solis et al., 2014)

An Antenna is an integral part of all devices that communicate through electromagnetic waves. For example, antennas are used in radio signal broadcasting, television broadcast, communication receiver, mobile communication, radar communication, and satellite communications. The antenna has a wide range of applications in the Bluetooth devices, wireless networks RFID based systems, and various other devices and systems. An antenna system normally has an arrangement of a metal conductor that is connected electrically to the receiver or transmitter. A varying electric field (current) creates varying magnetic fields. These time-varying fields radiated from transmitting antenna to the open space as a transverse em wave. (Aghwariya et al., 2020; Kalam et al., 2018; Solis et al., 2014)

At the reception of the em wave, the time-varying electric and magnetic field of the coming wave put force on the electrons of the antenna elements which create a varying current in the antenna.

## 1. INTRODUCTION OF THE WEARABLE ANTENNA

These types of antenna are different from the conventional microstrip antenna. The wearable antenna can be installed on the body or any cloth. Normally wearable antennas are installed on a piece of cloth for communication they have many applications like tracking and navigation. They can also be used for public safety. (Kalam et al., 2018) In the textile antennas the element which is radiation the electromagnetic signal is made up of the conductive textile materials such as Pure copper polyester, taffeta fabrics Zelt, Flectron, etc. while the substrate is made up of the nonconductive material such as silk, fleece or felt. Since the last few years, wearable antennas are widely used in body-centric wireless communication and it became a very important branch of mobile communication. For enhancing the body-centric communication, the IEEE 802.15 an Asian committee has been established to regulate application. The communication through the wearable antenna takes place within the range of private and body area networks. The major application of the wearable textile antenna is to communicate and track the person which is wearing the cloth on which the textile antenna is installed.

One of the leading research topics of the antenna technology is a wearable fabric-based antenna. It is also known as the wearable antenna. Wearable antenna technology has several advantages like low cost, lightweight, negligible maintenance, easy installation, and various others. (Aghwariya et al., 2020; Kalam et al., 2018) Most famous type of warble antenna is to watch ghazal GoPro cameras and even the night plus sensor which connects with the smartphone via Bluetooth and it can be installed in user shoes.

## 2. WEARABLE TEXTILE ANTENNAS

Textile antennas are having a large application in health care, safety, and protection, lifestyle, etc. Body. These antennas are capable of detecting, sensing processing, and energy harvesting they are the growing solution of the monitoring and tracking of the peoples. The cloths with an installed wearable antenna will be able to detect track and sense, they can communicate, will be useful in many important fields of communication (Paracha et al., 2019). The textile antennas are the most useful way to convert the cloths into effective communication systems. There are various properties of the wearable textile antenna like thin, lightweight, low-cost easy tuning, and better efficiency, robustness. Thus, planar antennas are the favorite type of antennas instead of having the lower achievable bandwidth, the planar antennas have the good coupling of the antenna with the circuit, antenna feed process and frequency matching circuit but the planar antennas might fail to integrate on the clothing because of the flexibility of the substrate material and the flexibility of the microstrip structure. The body-worn antenna generally radiates perpendicular to the ground plane and the planar surface protects the tissues of the body (Lee et al., 2012). There are several differences between the microstrip antennas and the wearable antenna, there is a special requirement to design the wearable antennas like planar structure, and conductive material which is flexible by which the ground and patch will be designed. The behavior of the antenna is depending upon the characteristics of the dielectric material. The bandwidth is determined by the thickness of the substrate and permittivity of the dielectric material. The efficiency of the antenna is mainly depending on patch conductivity and ground plane conductivity; hence conductivity should be as high as possible.

## 3. DESIGNING CHALLENGES OF WEARABLE ANTENNA

There are 2 types of challenges that occur while designing a Wearable antenna

### 3.1. Proximity to the Human Body

The wearable antenna installed on the human body which works as a lossy dielectric material for the em wave. According to the property of the lossy dielectric material the human body converts electric fields into the heat. Some part of the energy of the em wave is also absorbed by the human body (Naresh et al., 2017; Sharma & Singhal, 2013). So, when any antenna placed close to the human body it reduces the efficiency of the antenna by the large amount. As per the studies if the same antenna is placed on or closes to the human body it reduced the efficiency by 70%. This is one of the main issues associated with the designing of these antennae .as the performance of the antenna is reduced drastically near the human body.

### 3.2 Keeping Small Size

The wearable antennas must have as thin as possible. the large size of the antenna as not possible to install on ant type of cloth. There is no point in having a watch with big antennas installed on it. To have the compact design of any device the designer provides very little space to install antenna hence it is very necessary to have the size of the antenna very small (Sharma & Singhal, 2013).

## 4. THE MATERIAL USED FOR DESIGNING OF WEARABLE TEXTILE ANTENNA

The wearable antenna technology provides the possibilities of continuous monitoring, harvesting of energy, better communication and storage. The wireless body networks have eide applications in health care and defense systems. The basic requirements of the body antennas are the planar structure, flexible material on which the antenna will be installed, antenna behavior will be defined by the material used as substrate. permittivity of the dielectric material and the thickness of the substrate is the deciding factor of the wearable antenna (Sharma, 2018; Sharma & Singhal, 2013). Efficiency of the antenna depends upon permittivity of the dielectric material and the thickness of the substrate. The use of textile material on which the use of textile antennas is designed is important to get the desired communication through the antenna. There are several textile materials are available in the market that can be used with the antenna. Normal regular textiles materials are used as a substrate, however very less data is available regarding the em properties of the regular textile materials (Aghwariya et al., 2020).

The textile materials generally used now adays to design the wearable antennas have low dielectric constant which also reduce its surface loss and increase the impedance matching and overall bandwidth. The textile materials have the properties to exchange the water molecules with nearby which effects the em properties of the antenna. Also, the textile materials are having some specific properties like porous anisotropic, and compressible material which can change its thickness and density with low atmospheric pressure. hence it is important to know how these characteristics change the antenna behavior and what way the unwanted effects can be minimized.

## 5. IMPORTANT PROPERTIES OF TEXTILE MATERIAL

Currently, two main types of textile materials are available in the market that can be used as a substrate in the wearable antennas. .1 Natural textile materials 2. Manmade textile materials.

The manmade fibers are very similar to the polymers in context with the molecular structure and also considered as a subcategory of the synthetic fibers. The textile antennas are larger in comparison with the high dielectric (Singhal, 2013). Many wearable antennas features contribute to the complete design features of the antennas. These features are determined as planar fibrous resources, belongings of which mainly resolute by the belongings of the module fibers, and the construction of the fabric. The general behavior of the textile material is determined by the compactness of the fibers used, the air volume in the fiber also the pore size. Air permeability and thermal isolation is also an important aspect which decides the general behavior of the textile material (Lee et al., 2012). The low atmospheric pressure can change the thickness and density of the material because the used fabrics are flexible and easily compressible. Hence intrinsic fibers can be introduced as the main and important property of the fiber. The properties of fibers materials are drastically affected by the water molecules and the surrounding of the fiber material. All the properties can easily affect the antenna applications. So, it is very important to know how these properties are changing the performance and characteristics of the antenna, so that these unwanted effects can be removed. Some important characteristics on which antenna performance is dependent are discussed here:

### 5.1 The Dielectric Constant of the Fabric

The main important property of dielectric materials is the permittivity of the materials, ε, which is a complex quantity. The permittivity is calculated in relative value to the free space permittivity. Generally, the dielectric properties of the material are decided by the frequency, temperature in which the material is going to used roughness of the surface, purity of the material, and the moisture content present in the surface. As Dielectric permittivity is and complex quantity the real part knows as a dielectric constant which varies with the frequency and the ratio of the imaginary to the real part is called loss tangent of the material. The characterization of the dielectric material depends upon the electric field orientation.

In this given table loss tangent and relative permittivity of various textiles are given.(Ouyang & Chappell, 2008)

*Table 1.*

| Nonconductive Fabric | ε'r | Tanδ |
|---|---|---|
| Cordura® | 1.90 | 0.0098 |
| Cotton | 1.60 | 0.0400 |
| 100% Polyester | 1.90 | 0.0045 |
| Quartzel® Fabric | 1.95 | 0.0004 |

The electrical properties of flat fabrics are characterized by surface resistivity. This is the ratio of the DC voltage drop per unit length per unit width to the surface current. This is a characteristic of the material and has nothing to do with the configuration of the electrode used for measurement.

## 5.2 Substrate Materials for Wearable Antennas

While designing the antenna of the wearable device plays a crucial role. Selecting the proper substrate material is very important for the overall performance of the device. The substrate material should be a low loss material so the antenna efficiency can be improved when it is placed near to the body. Selecting the suitable substrate for the antenna is important to increase the antenna efficiency and when the antenna is installed on the body (Solis et al., 2014). Many studies suggested that a low loss conductive fabric is best suited for the better performance of the critical to the performance of a textile antenna. The substrate material should be flexible in nature. The dielectric properties of the textile material changed with each different type of material and operating frequency. Before designing the antenna, the textile material with the proper dielectric properties should be chosen. the textile material should have low relative permittivity.

## 5.3 Conductive Properties of Fabric

Fabrics that exhibit electrical conductivity are called conductive materials. Conductive fabrics can be made of woven metal strands or conductive yarns, due to the metal coating, in the manufacture of conductive textiles. People are very interested in semi-propelled textiles, which make ordinary textiles impermeable to carbon-based or metal-based powders. Various fabrics that have the properties of motionless electricity and protect it from em field variations, as well as other properties such as heat regulation, and anti-bacterial properties, have been extensively used in the last few years. However, it soon became clear that fabrics made with all conductive fibers like steel, silver, gold, copper have the potential by easing the addition of "soft" system networks into fabrics, thereby production them a certainty 'smart'. Unlike many textiles, the function of smart clothing is not inactive: They have the intelligence to sense and respond to touch, temperature, etc. Smart textiles materials can be joint with antennas, GPS, and other varying devices such as mobile phones without interfering characteristics of the fabric: the appearance, feel and behavior of the conductive yarn can all behave like traditional fabrics.

Fabric is usually used as an on-off switch to any circuit to perform the specific operation of any other external electronic device. To switch a connection needs to be built between two fabrics.

Restraining the widespread use of smart textiles is not an urgent issue of evaporation, durability, and the most important source of power (Jiang & Werner, 2016). To work, all electronic textiles (electronic textiles) need to be charged. Although many developments have been made in the miniaturization of batteries, they are still troublesome and need to be taken out for cleaning. Therefore, the expansion of many smart applications depends on the better power supplies.

## 5.4 Moisture Absorption Capabilities

The fiber continuously exchanges water molecules with the air, and always establishes a dynamic balance with the air temperature and air humidity. When textile fibers absorb water, the dielectric constant of water is higher than that of fabric, and they will partially and axially tighten the fabric, thereby affecting the dimensional stability of the antenna. It will also change the electromagnetic performance and increase the dielectric constant and loss.

Besides, some more issues are occurring due to the wetness of antenna due to rain or washing of the fabric and the same issues also may occur due when it absorbs the moisture from the skin, these effects also change characteristics of the antenna (Matthews & Pettitt, 2009).

The fabric antenna easily absorbs moisture, so it will change the antenna's resonance frequency and impedance bandwidth. Even when the antenna absorbs sweat from the human body it also changes the resonating frequency of the antenna.

The fiber continuously exchanges water particles with the air, and always establishes an active balance with air temperature and air moisture. When textile fibers absorb water, the dielectric constant of water is higher than that of fabric, and they will partially and axially tighten the fabric, thereby affecting the dimensional steadiness of the antenna (Naresh et al., 2017). It also changes the electromagnetic performance, and increase the dielectric constant and loss. Also, the skin absorbs moisture due to rain or washing clothes and even absorbs skin moisture from the antenna, which also absorbs wettability.

## 5.5 Thickness of the Dielectric Material

The antenna bandwidth and antenna efficiency are mainly dependent on the substrate dielectric constant and substrate thickness. The typical value of the substrate thickness h is usually in the range of $0.003\lambda \leq h\lambda 0.005\lambda$, where $\lambda$ is the wavelength. For a fixed relative permittivity, the thickness of the substrate can be selected to achieve the maximum value of the antenna. However, this value cannot improve the efficiency of the antenna. hence the thickness of the dielectric material compromised according to the bandwidth and efficiency of the antenna (Kiourti & Nikita, 2012).

Due to the narrow magnetic permittivity value range of textile materials, its thickness may vary greatly. This is mainly due to the bandwidth. The input impedance and the resonating frequency is responsible for the depth of the dielectric material. It plays an important role in designing the antenna.

## 5.6 Mechanical Deformations of Dielectric Material

The arc of the human body is a superposition of curves in any direction. Textile materials have excellent flexibility and elasticity and are very forms, which leads to changes in its electromagnetic properties for these surfaces. However, when the fabric is resolved to the fabric surface, it bends and distorts, which affects the antenna performance. As mentioned previously, with the bending of the dielectric fabric the permittivity and substrate thickness changes, thereby affecting the resonance frequency, mainly the antenna bandwidth (Kiourti & Nikita, 2012).

Also, the geometrical changes due to the elongation and/or compression of the textile components will reduce the geometric accuracy and shape that affects its performance by resonating frequency of the antenna.

Besides, from the designing perspective of the antenna, the springiness of the fabric is inconvenient because it makes accurate meaning, and cutting of the shape is difficult, and many materials overlap without wrinkles. Woven and non-woven fabrics are more stable than knitted fabrics, resulting in higher geometric accuracy of the frame. Generally, the thickness of the yarn or fiber component decides the accuracy of the antenna (Duan et al., 2012). For example, the conducting braiding accuracy can allow 0.15 mm that's why knits are not used as a dielectric of the antenna. However, they can be properly stabilized if manufactured with a stiffer textile material, like higher tensile strength fabric.

## 6. MANUFACTURING TECHNIQUES OF WEARABLE ANTENNA

The wearable antenna performance is basically dependent upon the manufacturing process and design of the antenna. The making process of the antenna must be precise should promise good similarity with the simulation results and design and should enhance the toughness and antenna repeatability. These types of antenna are to be designed on the fabric and goes through the different and tough operating conditions such as frequent changing of user location, changing weather conditions and sudden temperature changes (Jiang & Werner, 2016). In addition, wash ability of the fabric is also important in case of using the embroidered textile materials. Embroidered antennas are having a good matching with the industry requirement of the textile materials.

There are two types of manufacturing process are used for designing of the antennas.

Rigid Antennas: The production process of these type of antenna uses conventional process of designing the antenna by using the etching process. These types of the antenna are worn on the user's body and can be miniaturized for user's convenience. These types of antenna are designated in various size and structure. reducing the size of the antenna is main drawback of these types of the antenna (Jiang & Werner, 2016).

Flexible Textile Wearable Antenna: The Textile antenna designing techniques can be divided in two categories. In the first type of method Thin and uniform metal is attached on the nonconductive type of Textile material and in the second method use of textile yarns are stich over nonconductive textile materials.

## 7. REQUIREMENTS FOR DESIGNING A WEARABLE ANTENNA

The most important part of the wearable device is its antenna. The antenna of any wearable device decides the long period operation without disturbing the user day to day activity and any behavioral changes. To improve the battery life the device should be energy efficient. We know that RF transmission causes the maximum part of the overall battery consumption. To increase the battery life of the device the gain of the antenna should be high which will reduce the number of retransmissions of the signal and enhance the link budget. One more important aspect of the device which is related to the battery of the device is the form factor. The suitable value of the form factor will also improve the battery life of the device (Singhal, 2013). The thin sized and flexible antenna should be used so that the sensor of the device can be incorporated easily. when a person wears and wearable device or it a device located close to the human body, so the human body tissue which is lossy makes changes into the frequency response of the device, and its radiation efficiency is reduced. When the antenna is worn by a person, its frequency response starts changing as well as the radiation efficiency starts degrading.

To overcome this problem the body phantom should be incorporated in the antenna design procedure and the antenna should be kept as far as possible to the human body electromagnetically (Fang et al., 2011). If all the precautions are to be included in the designing of the antenna then the overall efficiency of the device can be maintained and it will enhance the battery life of the device.

To further improve the efficiency of the device reconfigurable antennas can be used. The optimal radiation pattern variations dependent upon the link which is to be formed. The Directional antennas give the best results and used to perform off-body links. for launching the surface waves the vertical polarization of the waves considered the best. Forming the in-body links is complicated because the propagation medium is the body that is lossy. therefore, the near field losses should be minimized for directing the radiation. If the device is proposed for many types of a link or if the device is located in a boundary of a person, which is going to lead to a very dynamic channel, then the antenna should satisfy the on the body and off body links of the communication. The SAR value of the antenna should be under 1.6W/kg

## CONCLUSION AND FUTURE SCOPE

A brief theory about the designing and analysis of wearable antenna is presented in this chapter. various designing aspects of the wearable antenna and textile materials are taken into account for the study. Important and necessary designing features of a textile antenna and important points which are to be considered before selecting the textile materials are discussed, several textile materials are retaken into account for the study. Because of the rapid growth of the wireless communication technologies wearable textile antenna is having a great future ahead and will be the future of wireless communication in the coming years (Aghwariya et al., 2020).

The antenna manufacturers are working on the field of wearable technology since the last decade. There is rapid growth we have seen in the last couple of years. still, the expansion of wearable technology is quite slow in comparison with the other popular technologies. This is mainly due to less technological enhancement and also inadequate market needs. Recent popular technologies are motivated by the easy connectivity and need for data. A large number of people want easy to used technology and they are ready can spent on the things that can enhance their lifestyle and quality of life. Easy communication and tracking devices and sensors are being famous day by day. Currently, almost every communication and tracking system is having wireless connectivity. The next updating for communication technology makes each communication and tracking truly wearable and to be installed directly into the cloths. To achieve these textile wearable antennas comes into the picture. Electronic devices are continuously decreasing in and developing technologies allows the textile materials to harvest the energy. RFID (radio frequency identification) is a suitable example of the textile antenna, RFID can easily install on fabric, it requires no battery and also its size is very compact. current research into flexible devices will also increase the requirement of flexible antennas. The main advantage f the textile antenna is that they can easily be integrated on the cloths and they do not require further assistance of the user for continuous and unaffected working. The textile antennas are also independent of the size limitations which can allow the increase in gain and bandwidth. This is phenomenon is quite useful where the frequencies are smaller and wavelength is high.

## ACKNOWLEDGMENT

Autor is thankful to the Department of Electronics and Communication Engineering THDC- Institute of Hydropower Engineering and Technology, Tehri, and the Department of Electronics and Communication Engineering, KIET Ghaziabad for providing the opportunity and necessary help to complete the study of wearable technology and warble antenna.

## REFERENCES

Aghwariya, M. K., Agarwal, T., & Sharma, R. (2020). Frequency Scanning SIW Leaky Wave Horn Antenna for Wireless Application. In *International Conference on Intelligent Computing and Smart Communication 2019. Algorithms for Intelligent Systems*. Springer. 10.1007/978-981-15-0633-8_79

Duan, Z., Guo, Y. X., Xue, R. F., Je, M., & Kwong, D. L. (2012). Differentially fed dual band implantable antenna for biomedical applications. *IEEE Transactions on Antennas and Propagation*, *60*(12), 5587–5595. doi:10.1109/TAP.2012.2209197

Fang, Q., Lee, S. Y., Parmana, H., Ghorbani, K., & Cosic, I. (2011). Developing a wireless implantable body sensor network in MICS band. *IEEE Transactions on Information Technology in Biomedicine*, *15*(4), 567–576. doi:10.1109/TITB.2011.2153865 PMID:21571615

Jiang, Z. H., & Werner, D. H. (2016). A compact and robust circularly polarized wearable antenna using an anisotropic metasurface. *IEEE/ACES International Conference on Wireless Information Technology and Systems and Applied Computational Electromagnetics (ACES)*. 10.1109/ROPACES.2016.7465320

Kiourti, A., & Nikita, K. S. (2012). Miniature scalp implantable antennas for telemetry in MICS and ISM bands: Design, safety considerations and link budget analysis. *IEEE Transactions on Antennas and Propagation*, *60*(8), 3568–3575. doi:10.1109/TAP.2012.2201078

Lee, S., Kim, U., Kwon, K., Seo, W., & Choi, J. (2012). Design of on-body antenna for wireless body area network. In *Advance Communication Technology* (pp. 300–303). ICACT.

Manna, S., Bose, T., & Bera, R. (2018). Wearable Antennas for Medical Application: A Review. In A. Kalam, S. Das, & K. Sharma (Eds.), *Advances in Electronics, Communication and Computing. Lecture Notes in Electrical Engineering* (Vol. 443). Springer. doi:10.1007/978-981-10-4765-7_13

Matthews, J. G. G., & Pettitt, G. (2009). Development of flexible, wearable antennas. *3rd European Conference on Antenna and Propagation*.

Naresh, B., Singh, V. K., & Bhargavi, V. (2017). Dual band RF energy harvester for wearable electronic technology. *2017 Third International Conference on Advances in Electrical, Electronics, Information, Communication and Bio-Informatics (AEEICB)*, 274-277. 10.1109/AEEICB.2017.7972428

Ouyang, Y., & Chappell, W. J. (2008, February). High Frequency Properties of Electro-Textiles for Wearable Antenna Applications. *IEEE Transactions on Antennas and Propagation*, *56*(2), 381–389. doi:10.1109/TAP.2007.915435

Paracha, K. N., Abdul Rahim, S. K., Soh, P. J., & Khalily, M. (2019). Wearable Antennas: A Review of Materials, Structures, and Innovative Features for Autonomous Communication and Sensing. *IEEE Access: Practical Innovations, Open Solutions*, *7*, 56694–56712. doi:10.1109/ACCESS.2019.2909146

Sharma, R. (2018). Prototype of Slotted Microstrip Patch Antenna for Multiband Application. *International Journal of Engineering & Technology, 7*(3), 1199-1201.

Sharma, R., & Singhal, P. K. (2013). *Miniaturized Patch Antenna Loaded with Metamaterial for Microwave Sensor Quantum Matter*. Academic Press.

Singhal, P. K. (2013). *Dual Director Microstrip Planar Yagi-Uda Antenna for X-Band* (Vol. 3). IJECCT.

Solis, A., Whittow, W. G., Alexandridis, A. A., & Vardaxoglou, J. C. (2014). Embroidery and Related Manufacturing Techniques for Wearable Antennas: Challenges and Opportunities. *Electronics (Basel), 3*(2), 314–338. doi:10.3390/electronics3020314

# Chapter 8
# Low Cost Compact Flexible Textile Antenna With Partial Ground

**Janabeg Loni**
*PK University, India*

**Anand Kumar Tripathi**
*PK University, India*

**Vinod Kumar Singh**
https://orcid.org/0000-0002-0671-0631
*SR Group of Institutions, India*

## ABSTRACT

*In this chapter, a low cost compact flexible textile antenna is presented that has low cost due to partial ground plane, and the anticipated antenna is suitable for different wireless applications. The proposed antenna is utilized as a line feed and simulated from 2GHz to 8GHz frequency. The impedance bandwidth of proposed antenna is 106.30% at -10 dB reflection coefficient. The anticipated design simulated with CST software to achieve directivity at resonant frequency 3.2 GHz is 3.609dBi and at 4.8 GHz is 4.519 dBi.*

## 1. INTRODUCTION

Wearable textile antenna has many applications particularly in human body related communications in which antenna can be built-in with the clothes or body etc. These type of antennas provide enhanced form of communication and improved living

DOI: 10.4018/978-1-7998-7611-3.ch008

quality of human life (Balanis, 2004 & Yadav, Singh, 2019 & Singh, Dhupkariya, Bangari, 2017 & Yadav et. al, 2020). There will be three main types of wearable antenna which can be put under these three categories. In first case, these type of systems can be worn directly on the body; in the second case, systems which can be directly integrated with clothing and in third case, they can be used as accessories.

With the advancement of the technologies, some technology provides double band or triple band configuration. But today, technology demand isn't limited to band but also cost, weight, flexibility and other parameters. Textile antenna design has multiple characteristics like low weight, more cost effective and availability (Singh, Ali, Ayub, Singh, 2014 & Yadav et.al, 2020 & Singh et. al, 2015 & Srivastava et.al, 2013 & Singh, et. al, 2016). In such antenna, there are many strategies that we apply for a larger bandwidth and gain such as by using partial ground, different kinds of patch and its dimensions and also different kinds of substrates. When the substrate's height is increasing, the efficiency is also increased, but at the same time surface waves is introduced which is not desirable (Kushwaha et. al, 2018 & Loni, Ayub, Singh, 2014 & Srivastava et.al, 2014 & Dhupkariya, Singh, 2015 & Singh et. al, 2015).

This paper is designed keeping in mind that it can be applicable for broad band application and it can be also fit for those applications where flexibility is major issue. So, this article proposed a broad band slotted patch antenna with partial ground structure having substrate material jeans is used that lies in Bluetooth, WLAN and Hiper LAN (Loni, Singh, 2015 & Xu Ling, 2012 & Khare, Singh, Saxena, 2020 & Rawat et. al, 2014 & Singh et. al, 2015). In this paper, for enhancing band width, circular shape wearable textile antenna is presented. The substrate used for proposed antenna is jeans ($\varepsilon_r = 1.7$). The antenna is simulated with help of Computer Simulation Technology (CST).

## 2. ANTENNA DESIGN

The lower part of antenna structure is partial and made by pasting copper tape on the lower surface of jeans substrate. The reduced ground is ultimately provides large bandwidth. The antenna is very compact and light weight with ground dimension of 15.5mm×50mm and substrate dimensions are 50mm×50mm. The substrate height (h) is selected 1.0 mm having dielectric constant of 1.7. Initially the inner circle slot has been cut to improve bandwidth and after that one square slot in upper side and one square slot in lower side provide expected better frequency range and large bandwidth. The feed line dimension is 2x16mm that plays very important role in enhancement of bandwidth also less copper is needed to manufacture (Singh et. al,

2015 & Srivastava et. al, 2015 & Singh, Kabeer Singh, Ali, 2013 & Singh et. al, 2014 & Dubey et. al, 2018.

*Figure 1. Top view of proposed antenna*

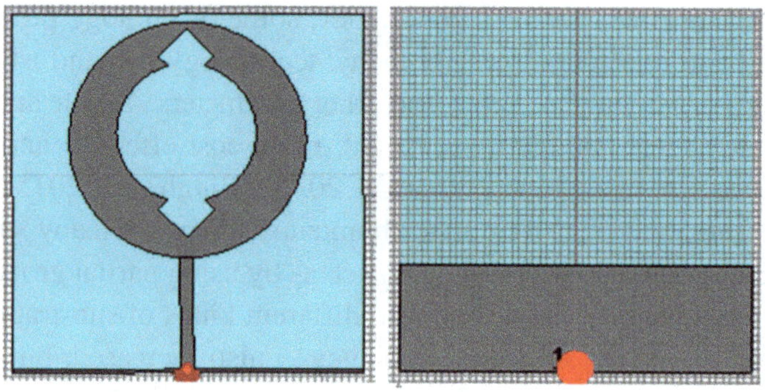

*Table 1. Antenna dimensions*

| Design parameters | Value |
|---|---|
| Width of ground plane ($W_g$) | 15.5 mm |
| Length of ground plane ($L_g$) | 50 mm |
| Width of substrate ($W_s$) | 50 mm |
| Length of substrate ($L_s$) | 50 mm |
| Substrate height (h) | 1.0 mm |
| Outer circle radius | 17 mm |
| Inner circle radius | 10 mm |
| Line Feed | 2×16 |

## 3. RESULTS AND DISCUSSIONS

The given antenna is cheaper and compact because of slotted structure that's why this can be used in small and light weight electronics devices. The magnitude of reflection coefficients is shown in figure 2. From figure 2, it is clear that the first corner frequency ($f_1$) is 2.445 GHz and the second corner frequency ($f_2$) is 8.0 GHz. From equation (1), the bandwidth can be calculated and finally the received bandwidth of given slotted antenna is 106.30%. Far field radiation pattern with phi=$0^0$ & phi=$90^0$ at 3.2 GHz and far field pattern with phi=$0^0$ & phi=$90^0$ at 4.8 GHz of anticipated antenna is shown in figure 3. And directivity is 3.609dBi at 3.2 GHz & 4.519 dBi at 4.8 GHz shown in figure 4. The simulated e-Field at 3.2 GHz and at 4.8GHz is shown in figure 5.

$$Bandwidth = \frac{f_2 - f_1}{\left(f_1 + f_2 \middle/ 2\right)} \times 100 \tag{1}$$

*Figure 2. Return Loss vs. Frequency plot of proposed microstrip antenna*

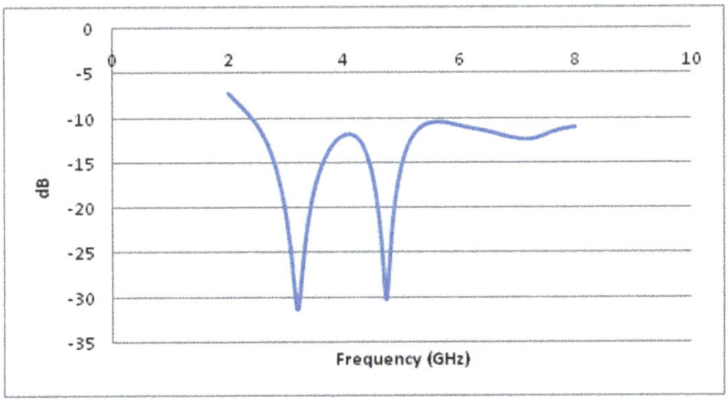


## 4. CONCLUSION

The proposed flexile antenna has been designed and simulated successfully for enhancing bandwidth. The anticipated antenna has been simulated in the frequency range of 2 to 8 GHz having bandwidth of 106.30%, and directivity is 3.609dBi at 3.2 GHz & 4.519 dBi at 4.8 GHz. The proposed antenna is simple, compact size and low cost as less copper will be needed to fabricate the patch and ground. Due to satisfactory performance of antenna it has many applications like in telecommunication, and different wireless communication systems.

*Figure 3a. 2D radiation pattern of anticipated antenna*

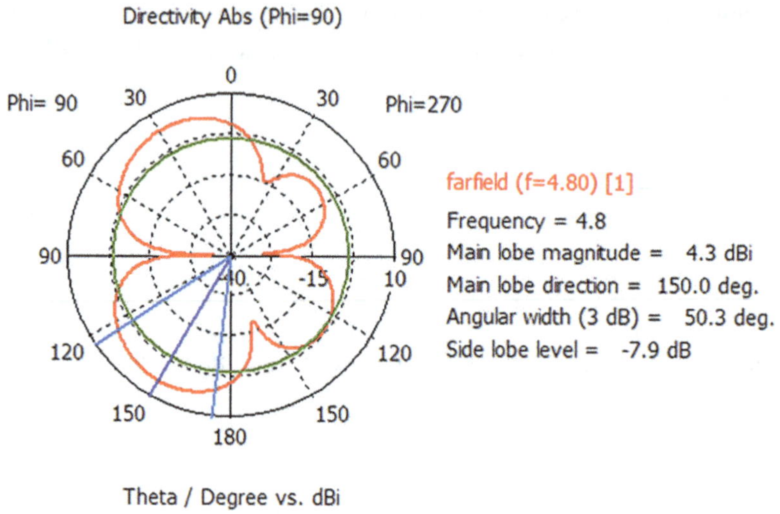


*Figure 3b. 2D radiation pattern of anticipated antenna*

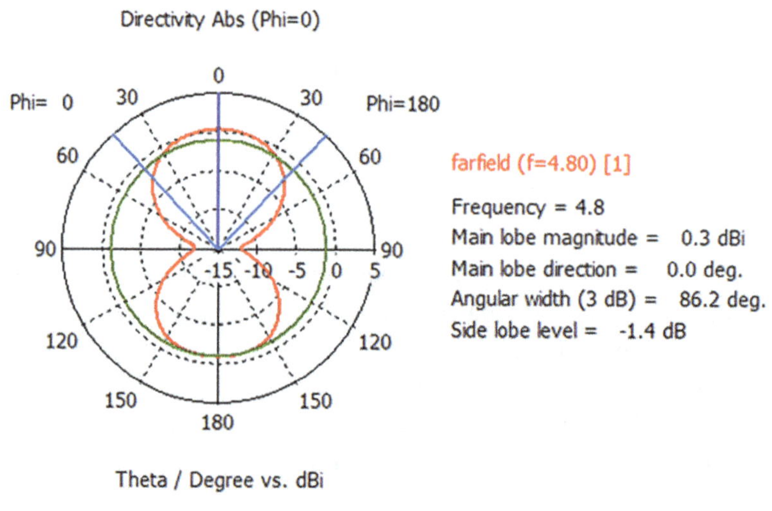

*Figure 3c. 2D radiation pattern of anticipated antenna*

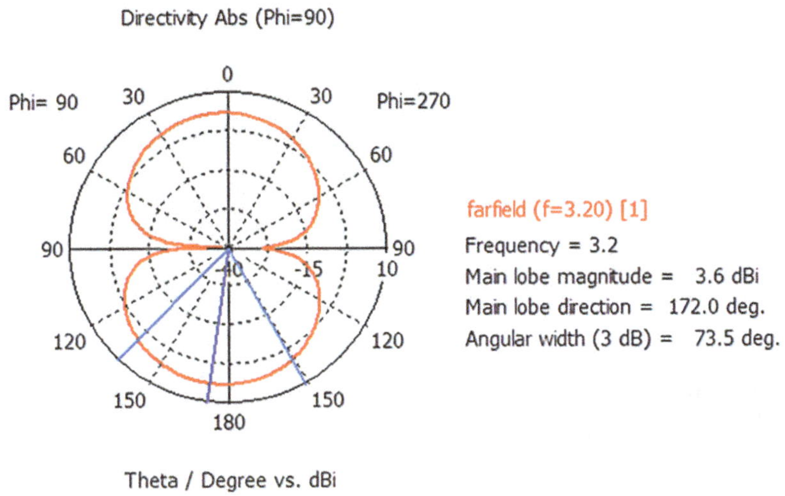


*Figure 3d. 2D radiation pattern of anticipated antenna*

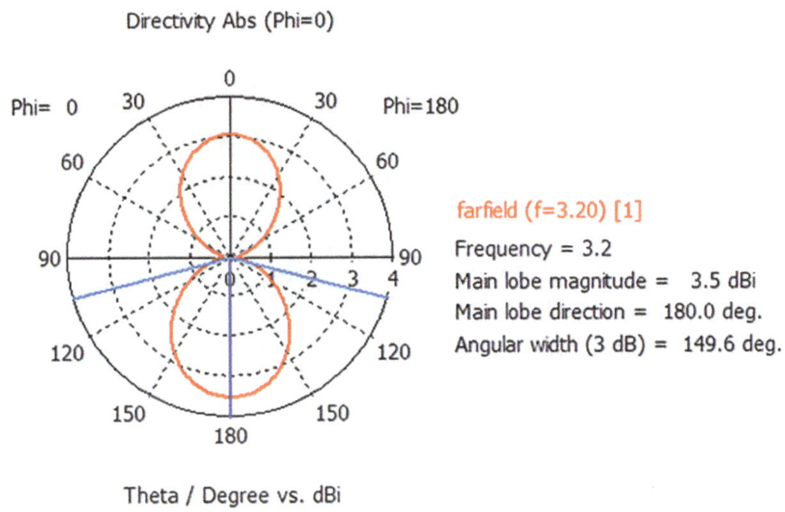

*Figure 4a. 3D radiation pattern for proposed textile antenna*

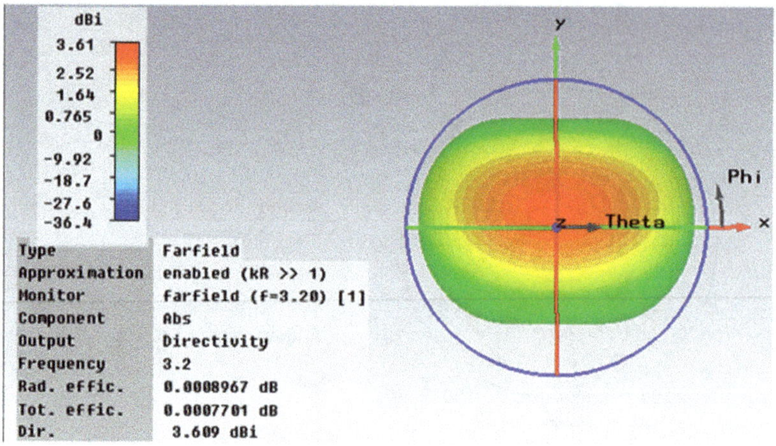


*Figure 4b. 3D radiation pattern for proposed textile antenna*

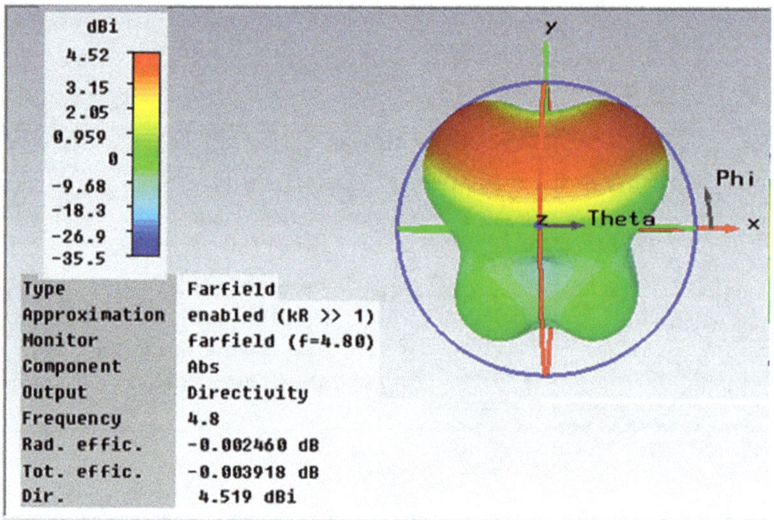

*Figure 5a. Simulated e-Field at 3.2 GHz*

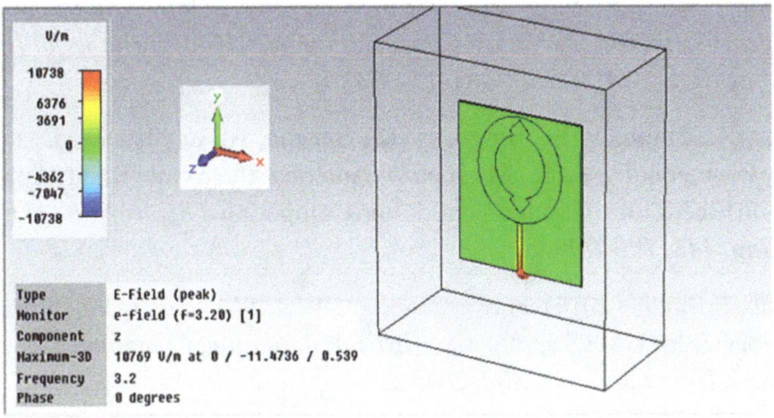


*Figure 5b. Simulated e-Field at 4.8GHz*

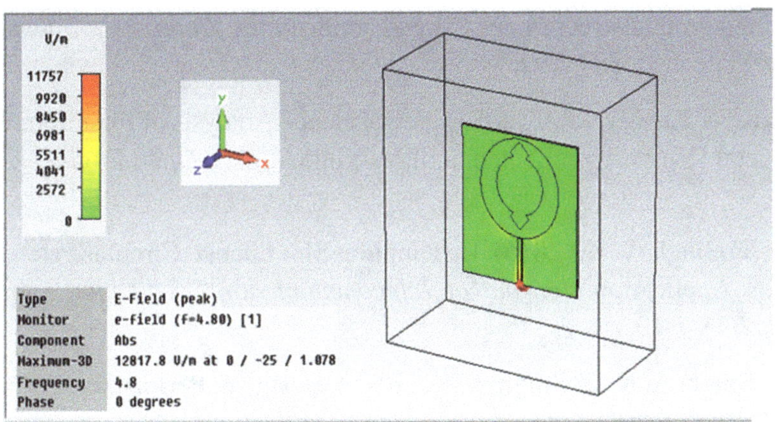


## REFERENCES

Balanis, C. A. (2004). *Antenna Theory: Analysis and Design*. John Wiley and Sons.

Dhupkariya, S., & Singh, V. K. (2015). Textile Antenna for C-Band Satellite Communication Application. *Journal of Telecommunication, Switching Systems and Networks, 2*(2), 20-25.

Dubey, R., Singh, V. K., Bhoi, A. K., & Ali, Z. (2018). Realization of trophy shaped flexible wearable antenna based on foam substrate. *International Journal of Engineering & Technology, 7*, 222-224.

Khare, B. B., Singh, V. K., & Saxena, A. (2020). Sigma-Structured Microstrip Antenna for Harvesting Energy for Low-Power Devices. In *International Conference on Intelligent Computing and Smart Communication 2019. Algorithms for Intelligent Systems*. Springer. 10.1007/978-981-15-0633-8_73

Kushwaha, R., Singh, V. K., Singh, N. K., Saxena, A., & Sharma, D. (2018). A Compact Pentagonal Textile Microstrip Antenna for Wide Band Application. Advances in Electronics, Communication and Computing. *Lecture Notes in Electrical Engineering*, *443*, 793–799.

Loni, Ayub, & Singh. (2014). Performance analysis of Microstrip Patch Antenna by varying slot size for UMTS application. *IEEE Publication, Communication Systems and Network Technologies (CSNT-2014).*

Loni, J., & Singh, V. K. (2015). Development of Bandwidth Enhanced Microstrip Patch Antenna for UMTS Application. *Journal of Microwave Engineering & Technologies, 2*(1), 1-7.

Rawat, S., & Sharma, K. K. (2014). A compact broadband microstrip patch antenna with defected ground structure for C-band applications. *Central European Journal of Engineering*, *4*(3), 287–292.

Singh, A. K., & Kabeer, R. A. (2013). Performance Analysis of First Iteration Koch Curve Fractal Log Periodic Antenna of Varying Angles. *Open Engineering, 3*(1), 51-57.

Singh, M., & Singh, V. K. (2015). Rectangular Slot Loaded Circular Patch Antenna for WLAN Application. *Journal of Telecommunication, Switching Systems and Networks, 2*(1), 7-10.

Singh, N., Singh, A. K., & Singh, V. K. (2015). Design & Performance of Wearable Ultra Wide Band Textile Antenna for Medical Applications. *Microwave and Optical Technology Letters, 57*(7), 1553-1557.

Singh, N. K., & Singh, V. K. (2016). Textile Antenna for Microwave Wireless Power Transmission. *Procedia Computer Science*.

Singh, V. K., Ali, Z., Ayub, S., & Singh, A. K. (2014). A wide band Compact Microstrip Antenna for GPS/DCS/PCS/WLAN Applications. In Intelligent Computing, Networking, and Informatics. Academic Press.

Singh, V. K., Dhupkariya, S., & Bangari, N. (2017). Wearable Ultra Wide Dual Band Flexible Textile Antenna for WiMax/WLAN Application. *International Journal of Wireless Personal Communications, 95*(2), 1075–1086.

Singh & Naresh. (2015). Multi Resonant Microstrip Antenna with Partial Ground for Radar Application. *Journal of Telecommunication, Switching Systems and Networks, 2*(1), 1-5.

Singh & Singh. (2015). Compact Circular Slotted Microstrip Antenna for Wireless Communication Systems. *Journal of Microwave Engineering & Technologies, 1*(1), 7-14.

Singh, Ali, Ayub, & Singh. (2014). Bandwidth Optimization of Compact Microstrip Antenna for PCS/DCS/Bluetooth Application. *Open Engineering, 4*(3), 281-286.

Srivastava, Ayub, & Singh. (2014). Dual Band Rectangular and Circular Slot Loaded Microstrip Antenna for WLAN/GPS/WiMax Applications. *IEEE Publication, Communication Systems and Network Technologies (CSNT-2014)*.

Srivastava, R., Singh, V. K., & Ayub, S. (2015). Comparative Analysis and Bandwidth Enhancement with Direct Coupled C Slotted Microstrip Antenna for Dual Wide Band Applications. In Frontiers of Intelligent Computing: Theory and Applications. Academic Press.

Srivastava, S., Singh, V. K., Ali, Z., & Singh, A. K. (2013). Duo Triangle Shaped Microstrip Patch Antenna Analysis for WiMAX lower band Application. *Procedia Technology, 10*, 554-563.

Xu, L., & Li, J. L.-W. (2012). A Dual band Microstrip Antenna for Wearable Application. *IEEE (ISAPE) Conference Proceeding*, 109-112.

Yadav & Singh. (n.d.). Design of U-shape with DGS circularly polarized wearable antenna on fabric substrate for WLAN and C-Band applications. *Journal of Computational Electronics*.

Yadav, A., Kumar Singh, V., Kumar Bhoi, A., Marques, G., Garcia-Zapirain, B., & de la Torre Díez, I. (2020). Wireless Body Area Networks: UWB Wearable Textile Antenna for Telemedicine and Mobile Health Systems. *Micromachines, 11*(6), E558.

Yadav, A., Singh, V.K., Yadav, P., Beliya, A.K., Bhoi, A.K., & Barsocchi, P. (2020). Design of Circularly Polarized Triple-Band Wearable Textile Antenna with Safe Low SAR for Human Health. *Electronics, 9*, 1366.

# Chapter 9
# Optimization of Textile Antennas Using Flexible Dielectric Material

**Deepa Rani**
*Sachdeva Institute of Technology, Mathura, India*

**S. K. Jaiswal**
*Modern College of Engineering, Jhansi, India*

**Vinod Kumar Singh**
https://orcid.org/0000-0002-0671-0631
*SR Group of Institutions, India*

**Nikhil Kumar Singh**
*Department of Education, Orai, India*

## ABSTRACT

*In this chapter, bendable textile antenna with partial ground plane is proposed. The anticipated antenna is made of jeans substrate which acts as a flexible material. The anticipated antenna is compared with some earlier published papers. The CST microwave studio is used to model the anticipated antenna, and after simulation, various results are obtained. The anticipated antenna has been provided the bandwidth from 1.90 GHz to 6.81 GHz having wide bandwidth of 112.74%. The designed antenna is suitable for wideband and Bluetooth application.*

DOI: 10.4018/978-1-7998-7611-3.ch009

## 1. INTRODUCTION

The pliable materials reduces the heaviness of antenna due to their low dielectric constant and is therefore responsible for the enhancement of the bandwidth. The antenna size varies inversely with the permittivity of the substrate that's why substrate of low permittivity is chosen for the fabrication of antenna (Bangari et al., 2018; Gao et al., 2017; Jain et al., 2019; Singh Nikhil & Singh Vinod, 2018; Wu et al., 2015). The characteristics of antenna e.g. the reflection coefficient, VSWR and radiation are investigated for suitable planning of antenna (Jamshed, 2015; Mourad & Essaaidi, 2014; Rachana et al., 2018; Rashmi et al., 2018; Sharma et al., 2018; Singh Nikhil et al., 2018).

The most of the existing antenna are not flexible and bendable and is identified as microstrip antenna. The lower and upper part of conventional textile antenna is made of copper (Abhishek et al., 2018; Grilo & Correra, 2015; Naresh & Singh, 2020; Poornima et al., 2018; Rahul & Singh, 2018; Rawat & Sharma, 2014; Xu & Li, 2012). Textile antenna is fabricated using copper tape and jeans is placed in between them. The development of textile antenna has been become easy because of miniaturization and usage of flexible materials (Jain et al., 2013; Kunal et al., 2018; Manju & Singh, 2018; Niharika et al., 2018; Pathak et al., 2012; Princi et al., 2018). When laptop, PC, and phone are in communication with the human body then the far field characteristics plays an important role. (Ashish et al., 2018; Deepak & Pathak, 2013; Jiang et al., 2014; Singh et al., 2010; Singh et al., 2013). The anticipated antenna acts as a receiver and can receive RF power at 2.163 GHz.

The anticipated bendable antenna is exclusive in terms of bandwidth and directivity The primary intention of designing a novel flexible antenna using a textile material for flexibility, compatibility and provide higher bandwidth that can be used for many applications such as Bluetooth, and other wireless applications The presented bendable textile antenna is applicable in broadband application in frequency range 1.90 GHz to 6.81 GHz.

## 2. DESIGN OF TEXTILE ANTENNA

The flexible antenna was recreated in CST microwave studio 2010 using Jeans substrate of the dimension 53.2mm×59.8mm. The patch comprised of three circles on the substrate in circle shape having radius 18 mm, 14mm and 8mm respectively. A strip line of length 2 mm & breadth 35 mm is fabricated inside the circle. The two annular rings are designed in such a manner that both rings could be interconnected by feed line. The sickle slot has been introduced to get broad band bandwidth. The CST software is used to model the anticipated antenna which is shown in Figure1 and the design parameters are given in Table.1(Singh et al., 2015; Singh et al., 2017; Yadav & Singh, 2019; Yadav et al.,). The anticipated antenna resonates at frequency 2.163 GHz.

*Figure 1. Proposed antenna geometries (a) Ground Plane (b) Proposed Antenna*

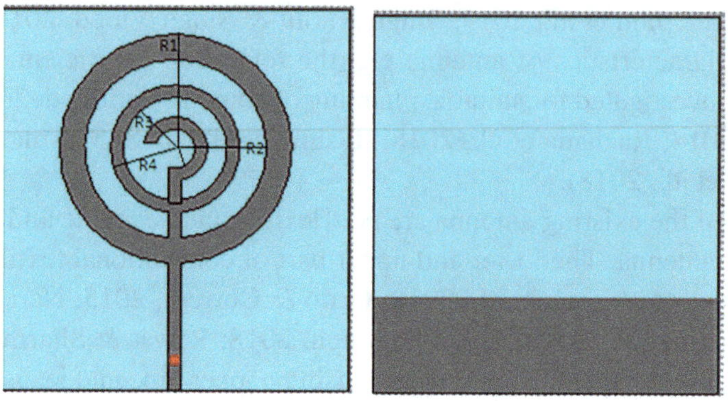


*Table 1. Textile Antenna Parameters*

| Antenna Parameters | Values |
|---|---|
| h | 1 |
| $\varepsilon_r$ | 1.7 |
| $R_1$ | 18 mm |
| $R_2$ | 14 mm |
| $R_3$ | 8 mm |
| $R_4$ | 10mm |
| Ls*Ws | 53.2*59.8 |

## 3. OPTIMIZATION OF PROPOSED MICROSTRIP ANTENNA

Three geometries of antennas are estimated to get optimum result as shown in Figure 2. Out of three designs, antenna-3 shows the better results as compared to earlier two antennas. The comparative results are shown in Figure 3 with different colours.

*Figure 2. Optimization of Bendable Microstrip Textile antenna (a) Initial design (b) Improved design (c) Anticipated Antenna*

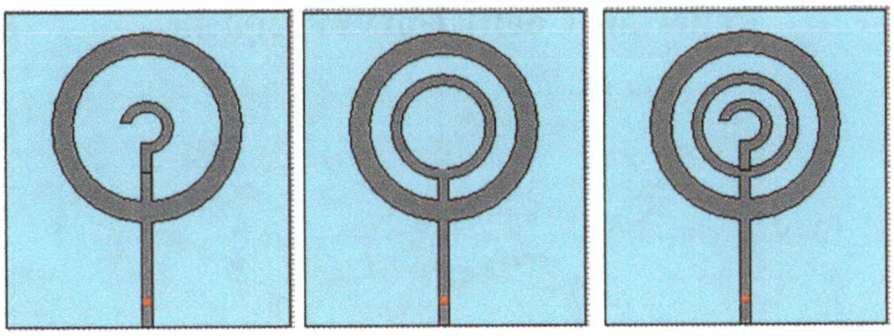

*Table 2. Comparative analysis with published papers*

| S. No. | References | Band (GHz) | Size (mm) | Band Width |
|---|---|---|---|---|
| 1 | Ling Xu (Xu & Li, 2012) | 2.4-2.5GHz | 70x40x3 | 2.40% |
| | | 5.725-5.875 GHz | | 6.40% |
| 2 | Marcus Grilo(Grilo & Correra, 2015) | 2.45 GHz | 43.3x38x1.4 | 15% |
| 3 | Rawat & Sharma (Rawat & Sharma, 2014) | 4.04-7.28 GHz | 30x30x1.59 | 60.30% |
| 4 | Anticipated Antenna | 1.90-6.81 GHz | 50x50x1.0 | 112.74% |

## 4. DISCUSSION

Fig. 3 shows the reflection coefficient at resonant frequency of 2.163 GHz. The far field 2-dimensional radiation pattern & 3-dimensional plot of bendable antenna is shown in Figure5 and Figure 6. Anticipated bendable antenna is better in terms of bandwidth and directivity as compared to some earlier published antennas which is revealed in table 2. (Hu et al., 2016; Loni et al., 2018; Saxena & Singh, 2018; Saxena et al., 2020; Singh & Singh, 2018; Yu et al., 2014).

*Figure 3. Reflection coefficient of different antennas*

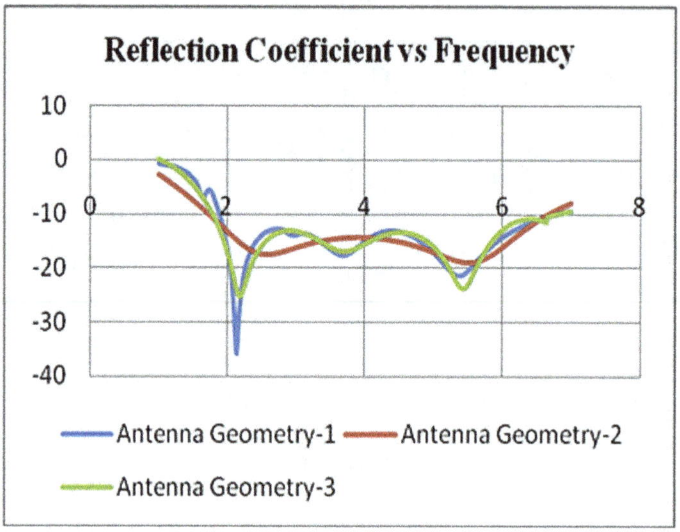


*Figure 4. Smith chart of bendable textile antenna*

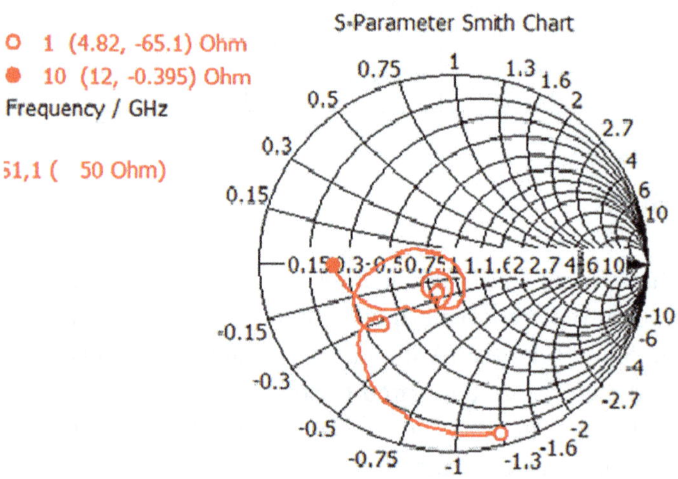


## 5. CONCLUSION

A novel design of bendable flexible textile antenna has been designed using textile material like jeans. The CST microwave studio is used to model the anticipated antenna and after simulation various results are obtained. The anticipated antenna has been provided the bandwidth from 1.90 GHz to 6.81 GHz having wide bandwidth of 112.74%.. The designed antenna is suitable for wideband and Bluetooth application.

*Figure 5. 2-dimensional radiation pattern of bendable antenna*

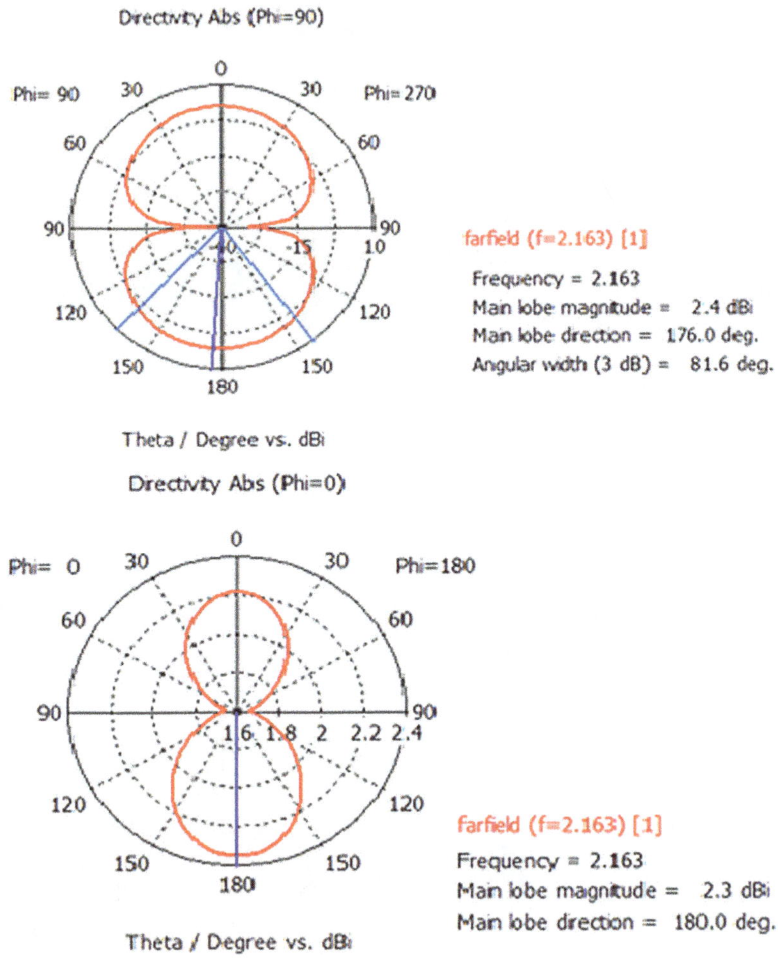

*Figure 6. Radiation pattern plot of bendable antenna*

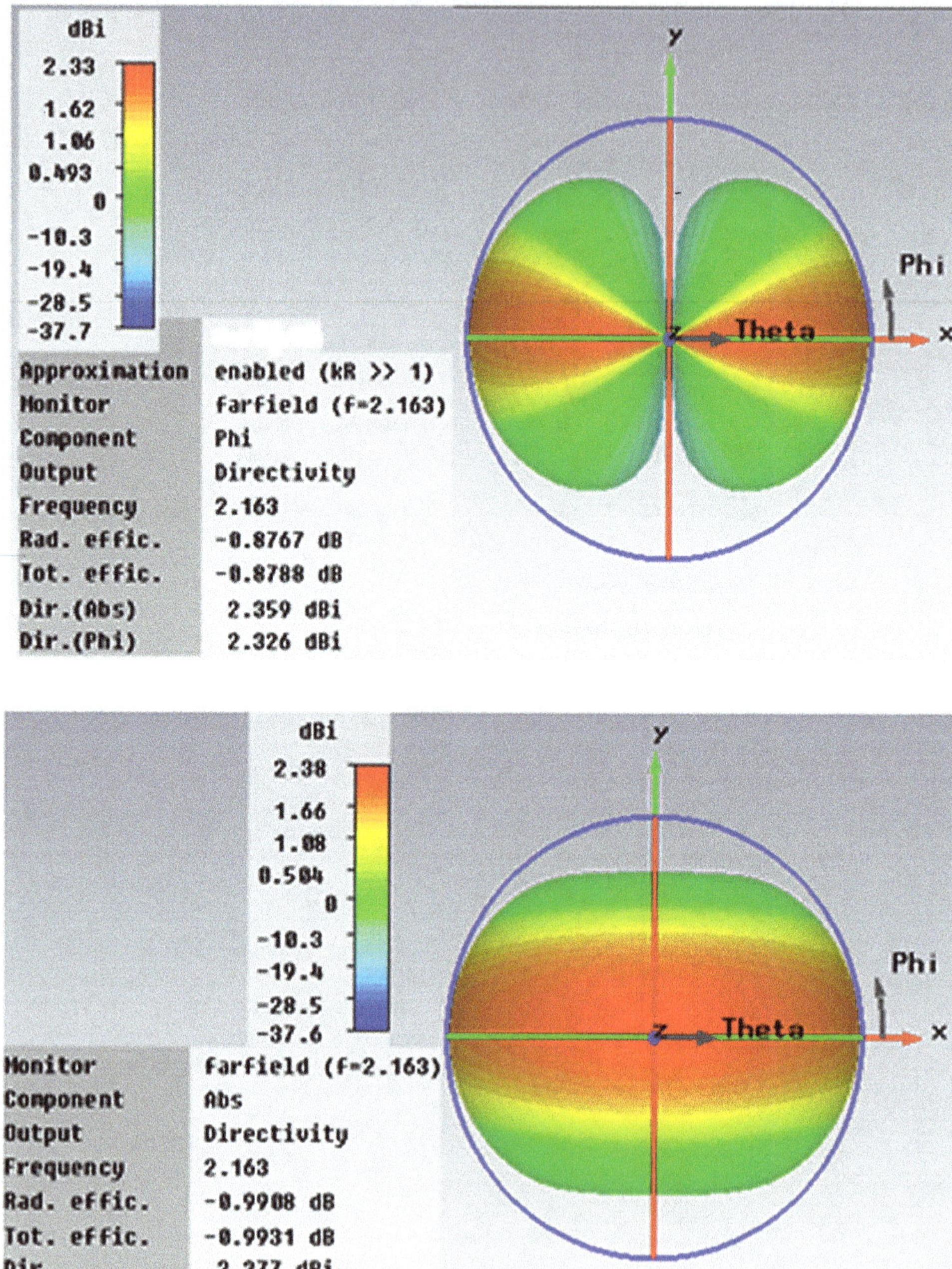

## REFERENCES

Abhishek, S., Sanjeev, G., Singh, V. K., & Bhoi, A. K. (2018). Design of Permanent Magnet Synchronous Generator for Wind Energy Conversion System. Lecture Notes in Electrical Engineering, 435, 23-32.

Ashish, K., Archana, L., Singh, V. K., & Bhoi, A. K. (2018). Estimation of Frequency Band of Microstrip Antenna (MSA) with Radial Basis Function (RBF). *Lecture Notes in Electrical Engineering*, *462*, 385–391. doi:10.1007/978-981-10-7901-6_42

Bangari, N., Singh, V. K., & Vanga, B. (2018). Dual-Band Wearable Rectenna for Low-Power RF Energy Harvesting. Lecture Notes in Electrical Engineering, 436, 13-21.

Deepak & Pathak. (2013). A study on inverted T shaped micro strip antenna at different frequencies. *International Journal of Engineering and Computer Science, 2*(11), 3180-3183.

Gao, G.-P., Hu, B., Tian, X.-L., Zhao, Q.-L., & Zhang, B.-T. (2017, April). Experimental study of a wearable aperture-coupled patch antenna for wireless body area network. *Microwave and Optical Technology Letters*, *59*(4), 761–766. doi:10.1002/mop.30408

Grilo, M., & Correra, F. S. (2015, July). Rectangular patch antenna on textile substrate fed by proximity coupling. *Journal of Microwaves, Optoelectronics and Electromagnetic Applications*, *14*, 103–112.

Hu, B., Gao, G. P., He, L. L., Cong, X. D., & Zhao, J. N. (2016). Bending andon-arm effects on a wearable antenna for 2.45 GHz body area network. *Antennas Wireless Propag IEEE Lett*, *15*, 378–381. doi:10.1109/LAWP.2015.2446512

Jain, A. K., Kanaujia, B. K., Dwari, S., Pandey, G. P., & Singh, D. K. (2019, June). Wideband circularly polarized magnetoelectric dipole antenna with I-slot for C-band applications. *Journal of Computational Electronics*, *18*(2), 660–670. doi:10.100710825-018-01295-y

Jain, S., Singh, V. K., & Ayub, S. (2013). Band Width and Gain Optimization of a Wide Band Gap Coupled Patch Antenna. *International Journal of Engineering Sciences & Research Technology*, 649-652.

Jamshed, A. (2015). A Novel Wide Band Microstrip-Line-Fed Antenna with Defected Ground for CP Operation. *Progress In Electromagnetics Research C*, *58*, 169–181. doi:10.2528/PIERC15052305

Jiang, Z. H., Brocker, D. E., Sieber, P. E., & Werner, D. H. (2014). A compact,low-profile metasurface-enabled antenna for wearable medicalbody-area network devices. *IEEE Transactions on Antennas and Propagation*, *62*(8), 402–4030. doi:10.1109/TAP.2014.2327650

Kunal, L., Archana, L., & Singh, V. K. (2018). Wide Band Triangular Patch Textile Antenna with Partial Ground Plane. *Advances in Intelligent Systems and Computing*, *632*, 577–583. doi:10.1007/978-981-10-5520-1_52

Loni, Singh, Tripathi, Chae, Sharma, & Bhoi. (2018). Improvement of the performance of wearable textile antenna. *IACSIT International Journal of Engineering and Technology*.

Manju, D., & Singh, V. K. (2018). Antenna for Wireless Area Network and Bluetooth Application. Lecture Notes in Electrical Engineering, 462, 377-384.

Mourad, M., & Essaaidi, M. (2014). A dual ultra wide band slotted antenna for C and X-band application. *Progress in Electromagnetics Research Letters*, *47*, 91–96. doi:10.2528/PIERL14052103

Naresh, B., & Singh, V. K. (2020). Integration of RF Rectenna with thin film solar cell to power wearable electronics. *International Journal of Microwave and Wireless Technologies*. doi:10.100740009-020-00877-4

Niharika, S., Singh, V. K., Rahul, S., & Prakash, S. J. (2018). Microstrip Textile Antenna with Jeans Substrate with Applications in S-Band. Lecture Notes in Electrical Engineering, 462, 369-376.

Pathak, R. S., Singh, V. K., & Ayub, S. (2012). Dual band Microstrip Antenna for GPS/ WLAN/WiMax Applications. *International Journal of Emerging Trends in Engineering and Development, 7*(2), 154-159.

Poornima, S., Singh, V. K., Archana, L., & Bhoi, A. K. (2018). Design and Analysis of Microstrip Antenna Using Multilayer Feed-Forward Back-Propagation Neural Network (MLPFFBP-ANN). *Lecture Notes in Electrical Engineering*, *462*, 393–398. doi:10.1007/978-981-10-7901-6_43

Princi, T., Upendra, T., Singh, V. K., & Bhoi, A. K. (2018). Solution of Economic Load Dispatch Problems through Moth Flame Optimization Algorithm. *Lecture Notes in Electrical Engineering*, *462*, 287–294. doi:10.1007/978-981-10-7901-6_31

Rachana, K., Singh, V. K., Singh, N. K., & Anurag, S. (2018). A Compact Pentagonal Textile Microstrip Antenna for Wide Band Application. Lecture Notes in Electrical Engineering, 443, 793-799.

Rahul, S., & Singh, V. K. (2018). Multi Resonant Textile Antenna with Partial Ground for Multiband Applications. Lecture Notes in Electrical Engineering, 462, 359-367.

Rashmi, S., Singh, V. K., & Puneet, K. (2018). A Compact CPW-Fed Defected Ground Microstrip Antenna for Ku Band Application. *Lecture Notes in Electrical Engineering*, *443*, 231–237. doi:10.1007/978-981-10-4765-7_24

Rawat, S., & Sharma, K. K. (2014). A compact broadband microstrip patch antenna with defected ground structure for C-band applications. *Central European Journal of Engineering*, *4*(3), 287–292.

Saxena & Singh. (2018). A Moon-Strip Line Antenna for Multi-Band Applications at 5.44 GHz Resonant Frequency. *2018 Fourth International Conference on Advances in Electrical, Electronics, Information, Communication and Bio-Informatics (AEEICB).*

Saxena, Singh, Lohia, & Faisal. (2020). A Jawbreaker-Shaped Fragmentary Ground Textile Flexible Antenna. *International Journal of Electronics, Communications, and Measurement Engineering*.

Sharma, Yadav, & Singh. (2018). Design of Circularly Polarized Antenna with Different Iterations for UWB Applications. Lecture Notes in Electrical Engineering, 443, 441-447.

Singh, Ali, & Ayub. (2010). Design of compact triple band microstrip antenna for wireless communication. *International Journal of Electronics & Communication Engineering, 3*(1), 323-330.

Singh, Ali, Singh, & Ayub. (2013). Dual Band Microstrip Antenna for UMTS/WLAN/ WIMAX Applications. *International Conference on Communication Systems and Network Technology (CSNT-2013),* 47 – 50.

Singh, N., Singh, A. K., & Singh, V. K. (2015). Design & Performance of Wearable Ultra Wide Band Textile Antenna for Medical Applications. *Microwave and Optical Technology Letters, 57*(7), 1553-1557. doi:10.1002/mop.29131

Singh, V. K., Dhupkariya, S., & Bangari, N. (2017). Wearable Ultra Wide Dual Band Flexible Textile Antenna for WiMax/WLAN Application. *International Journal of Wireless Personal Communications, 95*(2), 1075–1086. doi:10.100711277-016-3814-7

Singh, V. K., & Singh, N. K. (2018). Rectenna design for electromagnetic energy harvesting and wireless power transfer. *IACSIT International Journal of Engineering and Technology*.

Singh Nikhil, Niraj, Zakir, Singh, & Bhoi. (2018). Inset Fed Circular Microstrip Antenna with Defected Ground. Lecture Notes in Electrical Engineering, 443, 605-611.

Singh Nikhil, K., & Singh Vinod, K. (2018). A Compact Slotted Textile Patch Antenna for Ultra-wide Band Application. Lecture Notes in Electrical Engineering, 443, 53-59.

Wu, C.-M., Syu, J.-W., & Liu, W.-C. (2015). Dual-band slotted patch antenna with defective ground forWLAN/WiMAX applications. *Progress in Electromagnetics Research Letters*, *53*, 1–6. doi:10.2528/PIERL15022503

Xu & Li. (2012). A Dual band Microstrip Antenna for Wearable Application. *IEEE (ISAPE) Conference Proceeding*, 109-112.

Yadav & Singh. (2019). Design of U-shape with DGS circularly polarized wearable antenna on fabric substrate for WLAN and C-Band applications. *Journal of Computational Electronics, 18*(3), 1103–1109. . doi:10.100710825-019-01342-2

Yadav, A., Kumar Singh, V., Kumar Bhoi, A., Marques, G., Garcia-Zapirain, B., & de la Torre Díez, I. (2020). Wireless Body Area Networks: UWB Wearable Textile Antenna for Telemedicine and Mobile Health Systems. *Micromachines, 11*(6), E558. doi:10.3390/mi11060558

Yu, X. J., Tang, Q., Liang, M., & Xin, H. (2014). Study of power efficiency of non-foster impedance matching for electrically small antenna. *IEEE Conf Publ,* 130. 10.1109/USNC-URSI.2014.6955512

# Chapter 10
# Review on the Development of Solid State Transformer

**Bharat Bhushan Khare**
https://orcid.org/0000-0001-8755-9808
*Rajiv Gandhi Proudyogiki Vishwavidyalaya, India*

**Rajeev Shankar Pathak**
*Eshan College of Engineering, Agra, India*

**Sanjeev Sharma**
*New Horizon College of Engineering, India*

**Vinod Kumar Singh**
https://orcid.org/0000-0002-0671-0631
*SR Group of Institutions, India*

## ABSTRACT

*According to future renewable electric energy distribution and management (FREEDM) system, solid state transformers play an important role in smart grid technologies. They have several advantages over conventional transformers such as bi-directional power flow, light in weight, compact size, etc. They also compensate the environmental issues which are created due to transformer oil. Because of various advantages over traditional transformer, SST is preferred widely at the present time. So in this chapter, the various architectures, needs, and applications of solid state transformers are discussed. The global market of SST has continuously improved because it has several applications and benefits.*

DOI: 10.4018/978-1-7998-7611-3.ch010

## INTRODUCTION

Solid state transformer which has the same working principle as conventional transformer with bi-directional power flow, consisted of using several power electronic switching devices and a high frequency transformer (Ronan et al., 2002),(Qin & Kimball, 2009). SST also known by different name such as Intelligent universal transformer (IUT), power electronics transformer (PET) and Energy control centers (ECC) (She et al., 2013). The solid state transformer (SST), which has been regarded as one of the 10 most emerging technologies by Massachusetts Institute of Technology (MIT) (Bowers et al., 1980) in 2010, has gained increasing importance in the future power distribution system. Basically the transformer enables the efficient and long distance transmission (Electronicsforu, n.d.). To overcome the deficiencies of traditional transformer like heavy weight, larger size and current limiting problem, solid state transformer was first introduced by William McMurray in 1970 (McMurray, 1971), but it cannot compete as conventional transformer because it had less efficient as conventional occurred. Figure 1 shows the concept of first electronics transformer.

*Figure 1. Principle of Electronic Transformer*

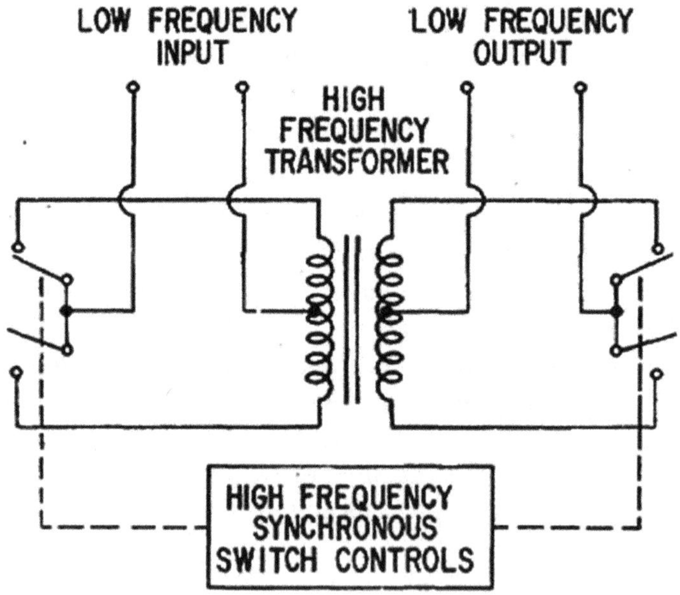


After the development of electronic transformer, the researchers trying to improve the result of solid state devices so that the quality and efficiency of such devices was improved and due to this the efficiency of the electronic transformer also improved. So in this era, in 1980, (Bowers et al., 1980) US Navy sponsored study was conducted to determine the feasibility of replacing electrical power transformer with solid state transformer by J.C. Brows et al.

In 1986, A.L. Goldberg et. al (Goldberg et al., 1987) described the issues related to the design of 1-10 MHz Transformer. They were conduct an analysis of skin and proximity effects in the conductors and measurements of permeability and hysteresis loss in the magnetic material with the computer-assisted study of the relationships between size, efficiency, and frequency.

Pawel M. Gradzki et al proposed a computer-aided tool for design and analysis of high frequency power transformer (Gradzki et al., 1990). This program used a multidimensional minimum search algorithm to design the smallest transformer and size of the magnetic core. The design specification includes current and voltage waveform, desired core shape and material, type of winding etc. the program also calculate the core loss, winding loss, skin effect and proximity effect. Mario Rabinowitz suggested the name of this device as (Rabinowitz, 2003) Compact Transformer.

The solid state transformer was introduced by William McMurray in 1970, but still there is a gap between 1970 to present time in the sense of commercialized application of SST. Because in 1970, this is not technical feasible (McMurray, 1971),(NITC, n.d.) and not comparable with conventional transformer in cost and efficiency. Efficiency is directly related to the ratio of thyristor blocking voltage to the voltage drop during conduction. So initially, it was used in special application (Bowers et al., 1980),(Energy, n.d.) where cost and efficiency are secondary to size and weight. The researchers worked to improve the quality of solid state device and formulate the efficient design of converters that can beneficial to increase the efficiency of SST, and now a day's some industries manufactured the SST and is also have the global market about some millions of dollar.

## NEED OF SST

The conventional transformer have some power quality issues i.e. if the distorted input is given to the primary side of traditional transformer then it results the output also is distorted and (Ronan et al., 2000),(Swathi, 2018),(Wang et al., 2007) the voltage drop under load, inability to mitigate flicker, environmental concern regarding to mineral oil, limited performance under dc offset load unbalances and a need for protection of the primary system, arises the problem inside or beyond the traditional transformer. The development trends of the traditional power transformer are mainly focused on new magnetic materials, manufacturing process and other economic factors (Prabhakaran et al., 2008),(Kunov, 2014). After some development, some complexities or difficulties are also exist in traditional transformer as heavy weight, large size, uncontrolled output, because when input of the primary coil changed then the output of the transformer is also changed. So due to these problem there is a need of some additional features that can able to fulfill such type of requirements.

Power electronic (Qin & Kimball, 2009),(Electronicsforu, n.d.) devices become advantageous technology to developmed the high power converter named as '' Solid State Transformer''. So SST is a device which behaves in the same manner (She et al., 2013) as a conventional power transformer in that is steps voltage level up or down while providing isolation and can accommodate load current of any power factor. The major advantages of this new approach are smaller size, less weight and controllable. It can also be able to suppress transients that may be present on the input line.

## CONSTRUCTION AND ITS ARCHITECTURE

In the basic construction of solid state transformer, it contains the (Dujic et al., 2013),(Manjrekar et al., 2000) power converter at both side as primary and secondary and high frequency transformer in the middle. Figure 2 describe the basic view of SST. For the high frequency transformer, the frequency lies in between few kilohertz to several mega-Hertz. To construct the core of HFT, nanocrystalline, ferrite, amorphous alloy is generally used.

*Figure 2. Basic construction of SST*

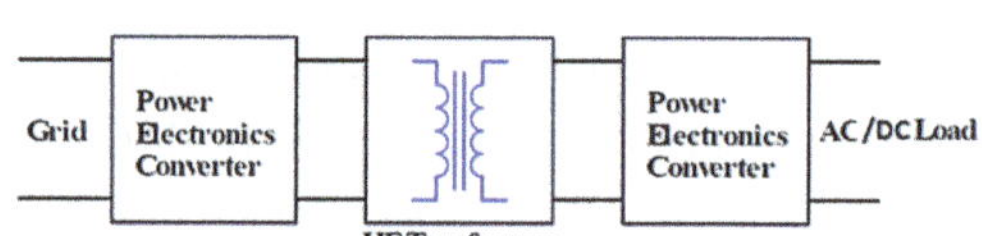


The architecture of SST can be sub categorized into three part such as single stage, double stage and three stage SST (She et al., 2013),(Bowers et al., 1980),(Swathi, 2018),(Manjrekar et al., 2000). In single stage the direct AC to AC conversion is performed and DC link is absent. In figure 3(a) single stage SST is shown . In Double stage SST, there may be either LVDC or HVDC link is to be present and its fully depends upon the requirement. According figure 3 (a), double stage SST contains only LVDC link so low voltage DC operated load can also be energized by using this architecture and in figure 3 (b), double stage SST using HVDC link is to be presented.

*Figure 3. Architecture of SST (a) single stage (b) Double stage with LVDC link (c) Double stage with HVDC link (d) Three stage with both LVDC & HVDC link*

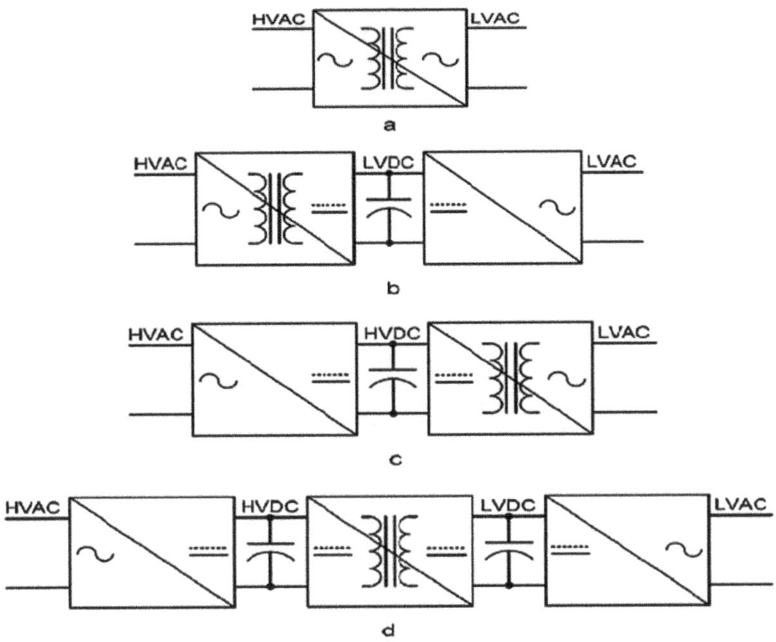


In three stage SST, the input supply converted into HVDC through rectifier then using isolated DC-DC converter (using high frequency transformer), it converted into LVDC and in the last stage it converted into LVAC so three types of output can be gathered through three stage SST which are HVDC, LVDC & LVAC. Figure 3 (d) shows the architecture of three stage SST.

## APPLICATIONS

SST have the virtue to gather the output AC as well as DC, so it have various advantages. SST can be used in different fields such as wind mill integration (Backlund et al., 2009),(Kolar & Ortiz, 2014), traction system (Komrska & Peroutka, 2008),(Ronan et al., 2002), distribution system (She & Huang, 2013), smart grid application. The distributed renewable electrical energy resources (DRER) and distributed energy storage devices (DSED) can be easily utilized by using solid state transformer. Figure 4 shows the virtue of SST which can easily handle DRER as well as DSED in micro-grid architecture.

*Figure 4. SST based micro-grid Architecture*

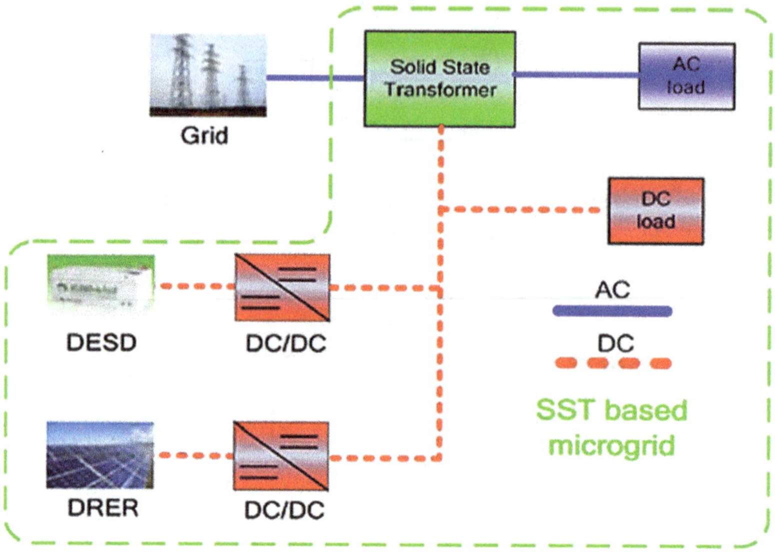


## CONCLUSION

SST plays an important role in all three basic components of power system i.e. generation, transmission & distribution. It can overcome the constraints of conventional transformer from the point of view of weight, size, harmonic suppression, current limiting. Researchers have proposed different topologies and architecture of SST to improve the efficiency and reducing the size moreover they have designed smart transformer by connecting real time monitoring system in SST.

## REFERENCES

Backlund, B., Rahimo, M., Klaka, K., & Siefken, J. (2009). Topologies, voltage ratings, and state of the art high power semiconductor devices for medium voltage wind energy system. *Proc. IEEE PEMWA*, 1–6.

Bowers, J. C., Garrett, S. J., Nienhaus, H. A., & Brooks, J. L. (1980, June). *A solid state transformer. In 1980 IEEE Power Electronics Specialists Conference*. IEEE.

Chen, Z., Guerrero, J. M., & Blaabjerg, F. (2006, June). A review of state of art power electronics for wind turbines. *IEEE Transactions on Industrial Electronics, 53*(4), 1002–1016.

Dujic, D., Zhao, C., Mester, A., Steinke, J. K., Weiss, M., Schmid, S. L., Chaudhuri, T., & Stefanutti, P. (2013, December). Power electronic traction transformer: Low voltage prototype. *IEEE Transactions on Power Electronics*, *28*(12), 5522–5534. doi:10.1109/TPEL.2013.2248756

Electronicsforu. (n.d.). https://electronicsforu.com/resources/learn-electronics/solid-state-transformers-power-grid

Energy. (n.d.). https://www.energy.gov/sites/prod/files/2017/09/f36/%5B5%5D%20NCSU%20%20Wensong%20Yu.pdf

Goldberg, A. F., Kassakian, J. G., & Schlecht, M. F. (1987, June). *Issues related to 1–10 MHz transformer design. In 1987 IEEE Power Electronics Specialists Conference*. IEEE.

Gradzki, P. M., Jovanovic, M. M., & Lee, F. C. (1990, March). Computer-aided design for high-frequency power transformers. In *Fifth Annual Proceedings on Applied Power Electronics Conference and Exposition* (pp. 336-343). IEEE. 10.1109/APEC.1990.66427

Kolar, J. W., & Ortiz, G. (2014, May). Solid-state-transformers: Key components of future traction and smart grid systems. *International Power Electronics Conference (IPEC), Hi-roshima.*

Komrska, T., & Peroutka, Z. (2008). Main traction converter with medium frequency transformer: Control of converters around MF transformer. *Proc. SPEEDAM*, 1194–1198. 10.1109/SPEEDHAM.2008.4581256

Kunov, G. (2014, May). Matlab-Simulink model of solid-state transformer realized with matrix converters. In *2014 18th International Symposium on Electrical Apparatus and Technologies (SIELA)* (pp. 1-4). IEEE. 10.1109/SIELA.2014.6871865

Manjrekar, M. D., Kieferndorf, R., & Venkataramanan, G. (2000, October). Power electronic transformers for utility applications. In *Conference Record of the 2000 IEEE Industry Applications Conference. Thirty-Fifth IAS Annual Meeting and World Conference on Industrial Applications of Electrical Energy (Cat. No. 00CH37129)* (Vol. 4, pp. 2496-2502). IEEE. 10.1109/IAS.2000.883173

McMurray, W. (1971). The thyristor electronic transformer: A power converter using a high-frequency link. *IEEE Transactions on Industry and General Applications*, *IGA-7*(4), 451–457. doi:10.1109/TIGA.1971.4181326

NITC. (n.d.). http://www.nitc.ac.in/electrical/ipg/pegcres/presentations/8.%20Dr.%20Ashok%20S/Dr%20ashok%20s%20.pdf

Prabhakaran, S., Stephens, C., Johnson, F., & Lorio, L. (2008). High frequency power transformers for future military propulsion and energy systems. *Proc. Electron. Mach. Technol. Symp.*, 1–21.

Qin, H. S., & Kimball, J. W. (2009). AC-AC dual active bridge converter for solid state transformer. *Proc. IEEE ECCE*, 3039–3044.

Rabinowitz, M. (2003). *Improved Power System of the Future.* arXiv preprint physics/0304083.

Ronan, E. R., Sudhoff, S. D., Glover, S. F., & Galloway, D. L. (2000, February). Application of power electronics to the distribution transformer. In *APEC 2000. Fifteenth Annual IEEE Applied Power Electronics Conference and Exposition (Cat. No. 00CH37058)* (Vol. 2, pp. 861-867). IEEE. 10.1109/APEC.2000.822606

Ronan, E. R., Sudhoff, S. D., Glover, S. F., & Galloway, D. L. (2002, April). A power electronic-based distribution transformer. *IEEE Transactions on Power Delivery*, *17*(2), 537–543. doi:10.1109/61.997934

She, X., & Huang, A. (2013, July). *Solid state transformer in the future smart electrical system. In 2013 IEEE Power & Energy Society General Meeting*. IEEE.

She, X., Huang, A. Q., & Burgos, R. (2013). Review of solid-state transformer technologies and their application in power distribution systems. *IEEE Journal of Emerging and Selected Topics in Power Electronics*, *1*(3), 186–198. doi:10.1109/JESTPE.2013.2277917

Swathi, G. (2018). Design an Efficient Power Electronics Transformer (PET) for the Improvement of Quality in Power. *IACSIT International Journal of Engineering and Technology*, *7*(3.12), 1046. doi:10.14419/ijet.v7i3.12.17629

Wang, D., Mao, C., & Li, J. (2007). Auto Balancing Electronic Power Transformer. *Zhongguo Dianji Gongcheng Xuebao*, *27*(6), 77–83.

# Chapter 11
# A Study of Smart Grid Systems

**Uliya Mitra**
*Bhilai Institute of Technology, India*

**Vikas Dubey**
https://orcid.org/0000-0002-0920-5593
*Bhilai Institute of Technology, India*

## ABSTRACT

*In the 21st century, energy supply has been one of the challenging issues that the world is facing. This is due to growing populations which results in more homes, businesses, and industries and has resulted in a large number of new appliances that increased the energy demand to a new level. The electric system we are using today is a one-way energy flow. Smart grid has emerged as the solution of continuously increasing demand. It is the future of electric energy system and also the power grid for upcoming generations. SG system works on the dual technique, that is, it counts electricity flow and the information which could deliver and distribute the energy universally by automating the existing electrical network.*

## 1. INTRODUCTION

The electric system of the world is and will be dominated by three factors. They are the needs of the consumer and its efficient supply, the new hardware and software technologies and lastly the government policies in every country to implement them. The conventional sources of energy are limited so now a days governments are switching towards the non-conventional or renewable sources of energy. This not only reduced the load on existing resources but also paved a way to save our environment, which is a concern for the whole world (Hamid Gharavi & Reza

DOI: 10.4018/978-1-7998-7611-3.ch011

Ghafurian 2011). Not only generation but the storage and conservation of generated power is also focussed. Since the life style of consumers results in a large amount of energy consumption so efficiency plays a major role here. Energy consumption on consumer side includes wide use of electric appliances in home, electric vehicles, usage of electricity in factories and industries. The simple methods to promote energy conservation includes remote control of in home appliances, the proprietorship for distribution of generated power from the inexhaustible energy resources can be done by managing the storage of electricity to match the energy supply and demand. Emerging technologies like sensors run by SCADA, intelligent autonomous controllers, secured 2-way communications and integrated data management, has opened up the opportunities to save energy that was not possible before twenty to thirty years. For the near future design of electric energy systems must satisfy this requirements by applying modern technologies for obtaining an acute, efficient and sustainable grid. For this a technique has been evolved named as Smart Grid. During recent years various articles published and conferences held about the smart grid. Here, I will define Smart Grid in its broadest global terms. For this I will be starting with the conventional electric energy system that is present and then identification of the areas will be done where changes are necessary for providing the intelligent and controlled method necessary for shifting to an efficient, shielded and secured future smart grid.

## 2. THE PREVALENT ELECTRICAL ENERGY NETWORK OR SYSTEM

The prevalent electricity delivery system has been represented in figure 1.

Basically, the system is broken into components such as: generation, transmission, substations, distribution, and the consumer end. For implementation of smart grid the following factors needs to be changed in conventional system:

1. The power generation which is centralized (J Katz 2008).
2. A uni-facial form of energy flow to customers from source side.
3. The pliant participation of clients as they have the limited knowledge of electrical energy usage which they gain by a monthly bill that is received at the end of a month. They should be more aware regarding the energy usage.
4. The monitoring in actual time and control has been restricted to power generation side and also in transmission side in few cases, distribution systems should also be added.

5. The electrical network should be pliable for injection of electricity at substitutive sources to any locus in grid, for aiding new amenities for systems more efficiently and sustainably.

*Figure 1. Conventional Electric Grid*
*(Roger et. al 2018)*

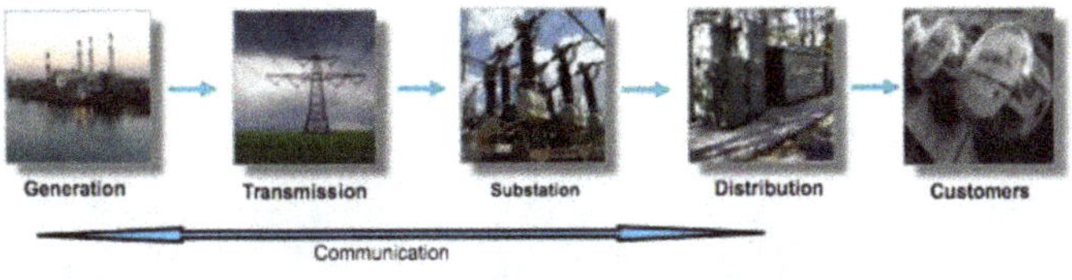


## 3. SMART GRID

Smart grid (SG) is an acute and upcoming electrical power grid, functioning within its own grid and also with other. It is an improvised version of the prevalent network for electric power. This updated grid network for electrical power allows flow of information on both way and same is the case for power shifting from suppliers end to consumers end because of the inculcation of management and communication systems for monitoring purpose. At starting days, idea of advanced metering infrastructure (AMI) was considered for operation of smart grid for improving energy efficiency which could be done by the construction of a grid which will protect and recover itself against malicious damage and natural disasters, also it was used for the improvisation of demand-side profile (Davood et. Al 2015). In existing grids, their main use the power transfer takes place from some of the centrally identified generators to various customers or users. On other side this SG system uses bilateral method of electrical energy flow and various information flow by making the system automotive and advanced in the field of energy delivery. With recent development in information technology and their adoption in Smart Grid, it will be easier to deliver power to varied conditions efficiently and responsibly.

Here Smart Grid refers to all those incidents occurring at any spot in the grid, like generation of power, its transmission and distribution, consumption and it also complies with the inline strategies already defined earlier (HE Brown & S Suryanarayanan 2009). One example is stated here to explain the process. Once a voltage transformer falling in medium range was failed being placed in distribution side of grid. Frequently reacting to the failure SG had changed the flow of power which helped in recovering delivery of power early. SG system is often considered as an electrical network or a system used for generating, transmitting, distributing and consuming electricity for a sustainable, effective, cyber secured system using approach of computation and mathematical employing technologies related to information gathering and communication process. Below in table 1 is shown the differences that exist between the prevalent grid system and modern Smart Grid system.

*Figure 2. The Smart Grid*
*(Roger et. al 2018)*

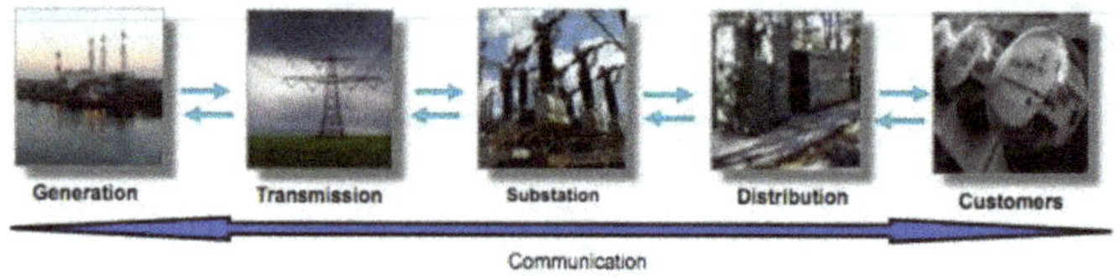


*Table 1. Differences between prevalent grid & Smart Grid*

| Existing Grid | Smart Grid |
|---|---|
| It is basically a Electromechanical system. | It is basically a digital system. |
| Unidirectional communication | Bidirectional communication |
| Generation of power is centralised | Generation of power is distributed |
| Sensor usage is less. | Throughout the system sensors are used. |
| Humans are required to monitor the system. | The system is self-monitored with the help of equipment's. |
| System is restored manually. | System heals by self. |
| Blackouts and failure occurrence is more. | It has the quality of islanding and being adaptive. |
| Control system here is a constraint. | Well managed and automatic control system. |
| System is limited to few customers. | In this system number of customer are more |

(Davood et al., 2015)

## 4. HISTORY OF SMART GRID

In 1896, on the basis of design of Nikola Tesla issued on 1888, commencement of first power grid was done (Xiaoyu Ma 2020). Since, last 50 years, the rapidity of electrical networks are unable to meet contemporary challenges, like threats related to security, national employment in power sector, proper distribution of power, profile of power quality demanded by customers. Keeping this disadvantages in mind the idea of Smart Grid was presented in 2005. Since, then it has been used continuously.

One of the largest and earliest grid had been installed by an Italian company named ENEL S.P.A, completed the installation project of SG in 2005. A project named Teleogestore made remarkable differences from the other existing systems in the consumer world. Integrated electronic meter that provides metering, contracted management and PLC communication functions were included in it. The Automatic Meter management having LV concentrators used to be placed on MV substations were communicated by telecommunication networks for public (like satellite, PSTN, GPRS & GSM) (A Pipattanasomporn and A. Rahman 2009). Bidirectional communication is managed by the LV Concentrators. Communication in Half-duplex form was done between digital meters and a central system metered remotely. This project saved 500 million euros from a capital of 2.1 billion euro invested annually and Smart Grid for the first time was commercially used.

## 5. NECESSITY AND BENEFITS OF SG SYSTEMS

The basic necessities of SG system will result in following outcomes:

- Emissions could be cut down by integrating renewable or non-conventional energy resources.
- Conservation of energy could be achieved more by participation of customers actively.
- Safety of systems could be ensured by employing communication processes that are cyber-secured.
- It can gain long term sustainability if the existing assets are used in a better manner.
- Optimized energy flow to reduce loss and energy price.
- To integrate electric vehicles which will reduce use of conventional fuels made up of hydrocarbons.
- Well managed distribution of generated power as well as their storage.
- Further integration of systems related to communication process and equipment employed for controlling energy system will promote the interoperability. Thus, improving and ensuring safe and flexible operation.

It is important to note that SG systems, as described previously, instead of replacing conventional or prevalent electrical network or system will build up the infrastructure already present which will increase the productivity of all the available assets encouraging the inculcation of newly introduced functionalities. The following characteristics of Smart Grid are not available in the conventional or existing systems of electrical energy network or system:

- A secured bidirectional communication which covers the whole system at each and every end point,
- Sensing of main components, checking, monitoring and detecting the variances in transformers, joints, cables etc. Also monitoring of power consumption by the consumer, quality of power and many more will be done in actual time.

All above mentioned characteristics will provide a massive amount of incoming data which could be converted into situational awareness for the state of the grid.

*Figure 3. The NIST conceptual model for smart grid*
*(Davood et al., 2015)*

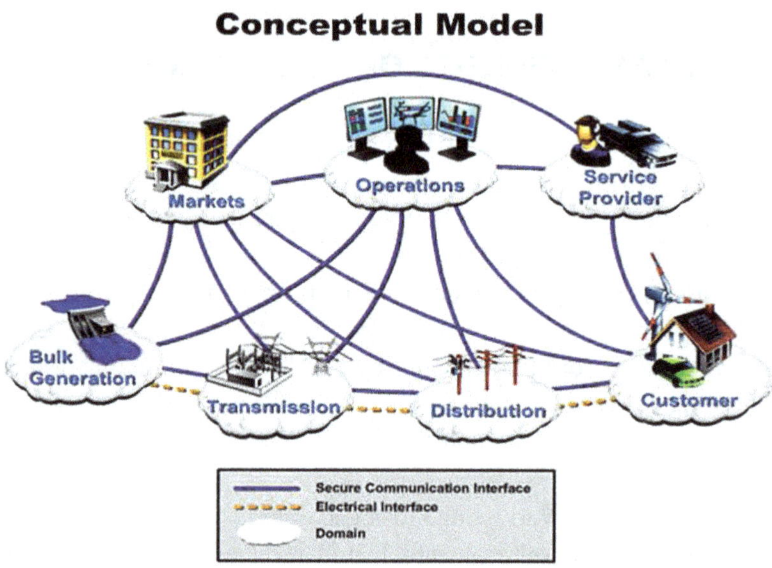


The controller technologies will automate and manage data and energy, so that information could be simplified, problems can be diagnosed at once, and corrective actions could be taken and performed instantly in the working field. Also the feedback loops are used to verify the work done with the help of metrics. Further the Smart Grid controllers will have the below mentioned features:

- Self-Recovery: The Smart Grid automatically repairs or removes the equipment having fault from the service before its failure and further reconfigures the system to ensure the supply of energy by proper route to all consumers for sustained powers.

- Flexibility: The fast, correct, safe and secure interconnection is done here for the distributed generation. Storage of energy of the system is possible at any spot and location(M Uslar et al 2010).
- Predictive: The controller in such system's machine could tell us the impact of weather earlier. Also they analyse the system stochastically to predict about upcoming events which could occur as a result they could protect or reconfigure the system in advance.
- Interactive: The information necessary about the status of the energy system are made available. All appropriate information are provided regarding the status of the system to both operators and the customers so that various candidates or utilities could participate in it so that they can actively optimize managing various outages or faults and contingencies.
- Optimisation: Real time status of every component is known. The control equipment provides optional routing paths for autonomously optimizing the flow of electricity all over the system or network.
- Security: Smart Grid having the capability of two-way communication covers each end of the system, as actual or realistic and security of cyber space is essential for all points and parts of the system.

Overall we can say that Smart Grid have a lot of benefits in terms of potential, economic and environment. They are as follows:

- Improved reliability on the quality and transmission of power(V Giordano & S Bossart 2012).
- Improved the efficiency of distribution of power. Also increased conservation of electrical energy.
- The costs to system for various utilities in the system are minimised as the expenses done by commercial and residential areas on use of electricity are reduced (Rahimi F & A Ipakchi 2010).
- Reduction in emission of all greenhouse gases.

*Figure 4. View of a smart grid showing working areas of control centres*
*(Zahran, 2013)*

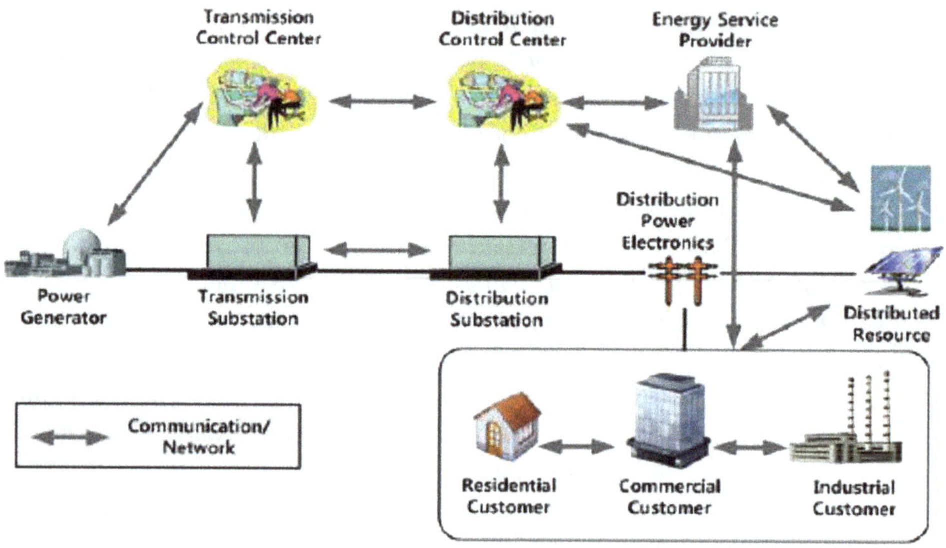


## 6. TECHNOLOGIES USED IN SMART GRID

1. AMI (Advanced metering infrastructure): A method for bidirectional metering and communication within utilities. AMI have two basic elements. Implementation of these two elements have been done. AMR or automatic meter reading systems is one among them. This proves to be an initial step towards the lowering of costs for data gathering by metering and using information of actual-time. MDM that is meter data management is the other element giving access to the integrated data of the meter's full range at one point. All the business process are automated by leveraging data for actual time. The data obtained is shared with lead business and other applications related to operation of the system for improving decision making ability and efficiency and for the whole company (T.Samad & A.M. Annaswamy 2011).
2. DMS (Distribution management system): It is a software modelling the distribution part of electrical network mathematically. It helps in predicting the effect of various outages in both transmitting and generating section, also it predicts the changes in voltage or frequency variation and many more. The capital required for investment is minimised because of better use of conventional assets. It also helps to reduce the capital investment by showing better utilisation of existing assets. It is also done by shaving the peak load by activating and observing the demand response (DR) (Indrajeet Prasad 2014).

3. The GIS (Geographic information system) is a technology designed and aiming mainly at the utilities industry for designing, modelling, also managing their special infrastructure. With the help of GIS, a graphical view of the infrastructure is obtained through simplified planning by integration of utility data which reduces cost.
    a. Outage management systems or OMS's are helpful in resolving outage related to speed which restores power more quickly and limits the outage costs.
    b. IEDs (Intelligent electronics devices) used here are advanced, critical and particular application active devices used for the program and are being installed in the field, they compute and helps in transmitting required information higher level (Yu Y Luan 2009). This devices further collect data from the meter both at network end and consumers end, further allowing the network reconfiguration.
    c. Next is the WAMS (Wide-area measurement systems) that provides systematic, inline and accurate measurements of power grids having vast range. A phasor measurement unit (PMUs) is further used to aggregate data and record events.
    d. At last EMS's (Energy management systems) (Mahmood Asif et al 2008) are used which at consumer end to control consumption of energy, generation of power, storage and charging of electric vehicles. This systems are been used in huge industries, places with commercial importance and further broad adoption of them with smart grids.

## 7. VISION OF THE FUTURE OF SMART GRID

The existing electric infrastructure will play a commendable role in upcoming days in the areas regarding power generation, lines for transmission of power, substation will be also included. But in future due to Smart Grid everyone will be connected to surplus, reasonable, structured and reliable electrical power at any instant and spot. Not only that, it would present a most secure electric service network. Being a fully and secured automated delivery network of power, it would have control over every customer and of each node by proper monitoring. It will ensure information as well as electricity flow within all power plants, appliances used by them and also between every node (Hamid Gharavi & Reza Ghafurian 2011). Due to technical advancements in superconductivity delivery of huge amount of electrical energy over long distances is possible now a days which also includes the congested areas with zero loss and a very low amount of voltage drop (Xiaoyu Ma 2020). Advancements in energy storage and the technologies for managing the demand side will eliminate the problems virtually that occur during the period when the load is at its peak. Since losses from outages for power will be low and improved quality of power due to minimum disturbances the customer will obtain the services regularly. The services will be reliable and will have minimum impact on environment. Smart Grid systems operate effervescently by opposing and defending every type of attack whether it is actual or cyber. The electric grid could achieve secured status if outages could be detected fast and the system could be restored as early as possible.

## 8. CHALLENGES FOR THE FUTURE SMART GRID

Smart Grid will improve and increase the capacity factor of the electric system with going time through demand management and more optimal supply (R Hassan & G Radman 2010). It would allow the re-use of existing hardware infrastructure in a more improved and efficient manner by adding a modern controller to the existing system. Evaluating the risks and impact on consumer regarding the use of available resources optimally would allow Smart Grid utilities to low down the peak demand and will also reduce the capital and O&M costs by reducing the emergencies of all kinds at the time of peak load periods. Thus it should be a joint task for the industries to maintain the tracking metrics around the whole world so that everyone could be benefitted.

Smart Grid in near future may face many challenges which could affect its success. Challenges such as consumer buy-in (Roger et al.2018). The consumers should be concerned regarding real savings and improvements in efficiency; regulation should be good. Also the governmental control must be up to date technologically and should connect all consumers; the cost should be justified; all the Smart Grid components should also be cost effective not only individually but systematically also (A Ipakchi, F Albuyeh 2009). In Education part: all utilities, service companies and the universities must produce educated consumers already accustomed and having good hands in computer sciences and systems engineering.

## 9. CONCLUSION

As we can see with the growing years it is the need to maintain our systems smartly. They will support the existing system and will improve the power system networks. As we have seen in this paper Smart Grid systems are really a boon for the existing systems. As they will maintain a two way process between the utility end to the consumer end. SG system will reduce the losses of the line as a result efficiency will be improved, distribute the power, reduce the emission, have a good control on the consumption, proper communication at each end and whole electric system. It will also make the consumers aware of the electric energy systems. So all nations of the world are working for making their electric systems smart so that electric energy supply should be done properly and reliably amidst of increasing demand.

## REFERENCES


Anderson, Ghafurian, & Gharavi. (2018). *Smart Grid: The Future of the Electric Energy System*. Research Gate.

Asif, M., Muhammad, A., & Irfan, A. M. (2008). Design and implementation of AMR Smart Grid System. *IEEE Canada Electric Power Conference*.

Brown, H. E., & Suryanarayanan, S. (2009). A survey seeking a definition of a smart distribution system. *North American power symposium'09*, 1–7. 10.1109/NAPS.2009.5484078

Gharavi, H., & Ghafurian, R. (2011). Smart Grid: The Electric Energy System of the Future. *Proceedings of the IEEE, 9*.

Giordano, V., & Bossart, S. (2012). *Accessing smart grid benefits and impacts: EU and U.S initiatives*. Academic Press.

Hassan, R., & Radman, G. (2010). Survey on smart grid. *IEEE Southeast on 2010*, 210–213.

Indrajeet. (2014). Smart Grid Technology: Application and Control. *International Journal of Advanced Research in Electrical, Electronics and Instrumentation Engineering, 3*(5).

Ipakchi, A., & Albuyeh, F. (2009). Grid of the future. *IEEE Power & Energy Magazine*, *7*(2), 52–62. doi:10.1109/MPE.2008.931384

Katz, J. (2008). Educating the Smart Grid. *IEEE Energy 2030*.

Ma, X. (2020). *Smart Grid*. https://www.cse.wustl.edu/~jain/cse574-10/ftp/grid/

Pipattanasomporn, A., & Rahman, A. (2009). Multi-Agent Systems in a Distributed Smart Grid: Design and Implementation. *Proc. IEEE PES 2009 Power Systems Conference and Exposition*. 10.1109/PSCE.2009.4840087

Rahimi, F., & Ipakchi, A. (2010). Demand response as a market resource under the smart grid paradigm. *IEEE Transactions on Smart Grid*, *1*(1), 82–88. doi:10.1109/TSG.2010.2045906

Samad, T., & Annaswamy, A. M. (2011). *The Impact of control technology- Control for renewable energy and Smart Grid.* www.ieeecss.org

Souran, Safa, Moghadam, Ghasempour, & Heravi. (2015). *Smart Grid Technology in Power Systems*. Research Gate.

Uslar, M., Rohjansand, S., Bleiker, R., Gonz'alez, J., Specht, M., Suding, T., & Weidelt, T. (2010). Survey of smart grid standardization studies and recommendations—Part 2. *IEEE PES*, *10*, 1–6. doi:10.1109/ISGTEUROPE.2010.5638886

Yu, Y., & Luan, W. (2009). Smart grid and its implementations. *CSEE*, *29*(34), 1–8.

Zahran, M. (2013). *Smart grid technology, vision management and control. WSEAS Transactions on Systems*, 12.

# Chapter 12
# Analysis and Design of Solar Dryer for Forest Produce of Bastar Region in State of Chhattisgarh, India

**Vikrant Tapas**
*Department of Mechanical Engineering, NMDC DAV Polytechnic, Dantewada, India*

**Rituraj Chandrakar**
*Department of Mechanical Engineering, NMDC DAV Polytechnic, Dantewada, India*

**Anil Kumar**
*Department of Mechanical Engineering, BIT, Durg, India*

**Prem Shankar Sahu**
*Department of Mechanical Engineering, BIT, Durg, India*

**Haldhar Ram Chandrakar**
*Department of Mechanical Engineering, CSIT, Durg, India*

**Yatendra Vinod**
*Department of Mechanical Engineering, NMDC DAV Polytechnic, Dantewada, India*

## ABSTRACT

*In the Chhattisgarh state of India, the Bastar region is famous for its various forest produce collected by local inhabitants. These collected forest produce are dried in open sun drying. This chapter presents the design and analysis of solar dryer which is simple in construction and a low-cost dryer. The construction of the solar dryer will be made of galvanized mild steel sheet absorbing the sun's radiation. The hot air will be taken out through a wind turbine ventilator at the top to maintain the required humidity level in the solar dryer. The analysis shows that the temperature inside the solar dryer is evenly distributed. The suggested solar dryer will reduce the drying time of the forest produce.*

DOI: 10.4018/978-1-7998-7611-3.ch012

## INTRODUCTION

As per FSI (Forest Survey of India) the state of Chhattisgarh is divided in the main three agro-climatic zones (Forest Survey of India, 2011). They are the Chhattisgarh plains, Northern hills of Chhattisgarh and the Bastar plateau. The Bastar plateau is divided in 7 districts. They are as Kanker (North bastar), Koandagaon, Narayanpur, Jagdalpur (Bastar), Dantewada (South Bastar), Sukma and Bijapur. This area is having large forest cover. The forest in this region include variety of forest that is very dense, moderately dense and open forest scrub. The local inhabitance collect the forest produce as per the season .There are nearly 41 forest produce is collected shown in Table 1(District Forest Office Dantewada, n.d.). The many of useful part of this collected forest produce is dried in open sun and then traded. The different part of forest produce mainly are fruit, seed, food pulp, flower seed, the hole part (accept root), leaf, rhizome, hole part (including root), root and bulb(Tapas & Sharma, 2014). In the shown Table 1 twenty eight items are collected on a large quantity and perfectly giving opportunity for solar drying.

## IMPORTANCE OF SOLAR DRYING

These forest produce is collected and dried in open sun. The open sun drying (OSD) is a traditional method having disadvantage as delayed drying, contamination by birds, insects etc, less hygienic, less clean, inferior quality product, uneven drying, more nutrient loss, space required is higher, low profit margins, drying possible only on sunny days, poor sensory qualities to products – appearance /color and textures and above all it is certainly not a good manufacturing practice (Sharma et al., 2009). The use of solar dryer improves the open sun in five important ways such as faster rate, more efficiency, more hygienic and healthier product (Aware & Thorat, 2012).

## CRITERIA OF SOLAR DRYER FOR BASTAR REGION

For the selection of solar dryer the main selection criteria are (Visvale, 2012):-

- Ease of design
- Can handle small and medium quantity of forest produce
- Assemble with easily available materials
- No use of electricity
- Less Cost

The selected design is further modified on the Ooi's suggested model (Ooi, 2006).It is basically a box dryer. To control the hot air flow inside the dryer a roof top turbine ventilator is used. The dryer is further analyzed and found the even temperature and velocity distribution in the dryer to maintain the desired humidity level is achieved through this. This will further improve the texture and quality of the forest produce. The Ooi (2006) incorporated solar drying into a box dryer with area 2.4 m X 2.4 m for bed depth of 0.3 m. This box dryer outer surface is covered by corrugated sheets. The selected dimension for analysis on Ooi's box dryer are kept 6 feet (1828.80 mm) in length and width is 3 feet (914.40 mm) and the upper height is kept as 2.62 feet (800mm) and lower height is 1.53 feet (467.19 mm) at maximum[fig. 1] . This will reduce the dimensions of Ooi's suggested model. The slant of the solar dryer will be at an angle of 170°. Figure shows the wire frame diagram of the suggested Solar Dryer. For the study the longitude is taken as 81.39909° and latitude is taken as 18.9739° [Fig. 3]. These coordinates are of Dantewada city.

*Figure 1.*

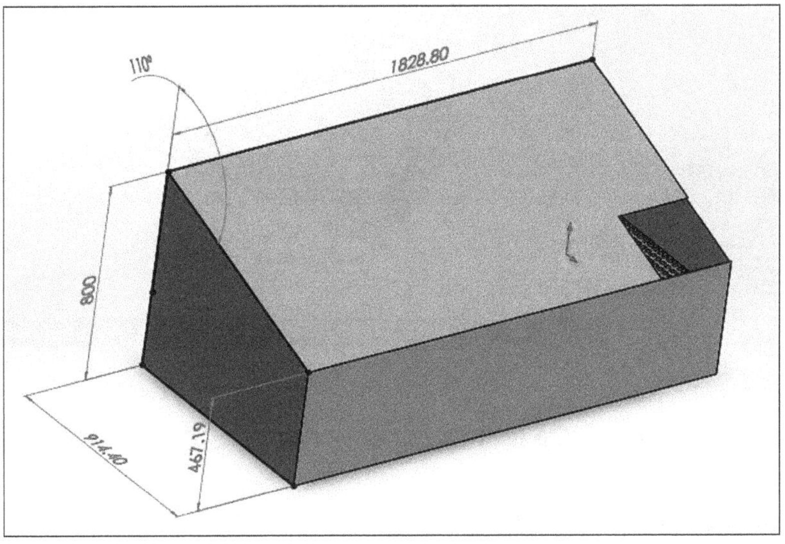

*Figure 2.*

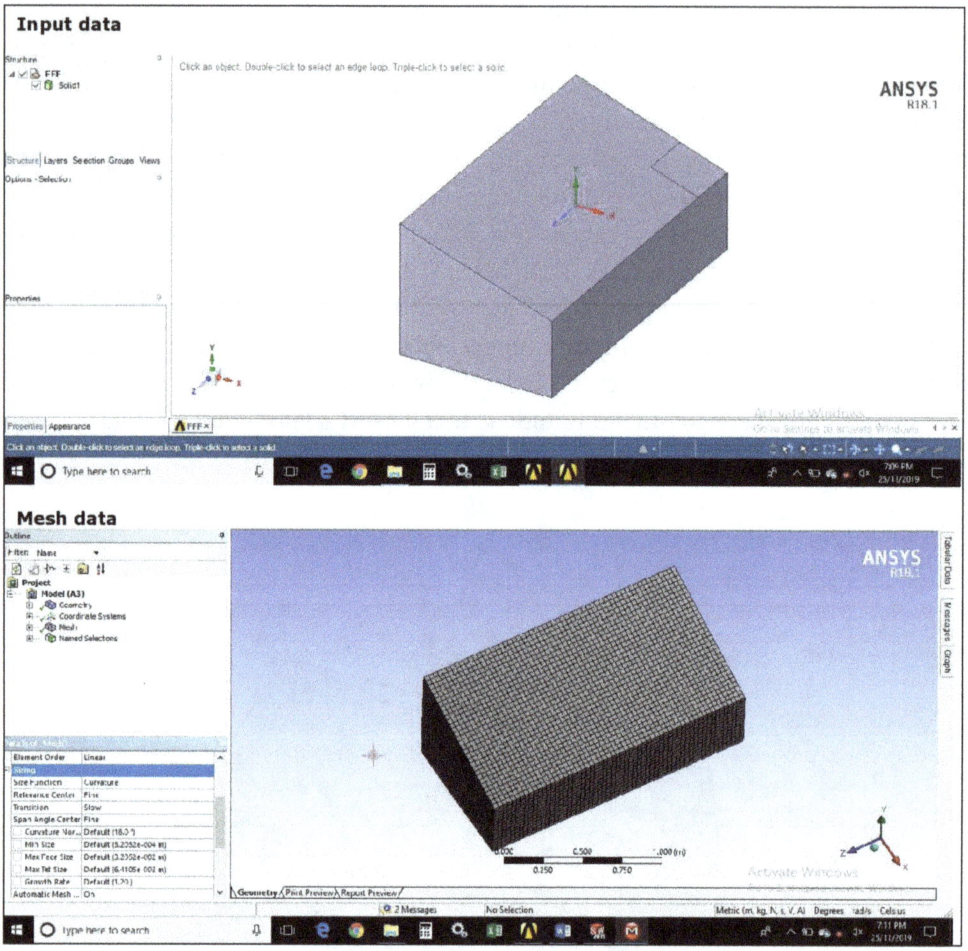


## CONSTRUCTION DETAILS OF PROPOSED SOLAR DRYER

The box dryer 6 feet (1828.8 mm) x 3 feet (914.4 mm) x 2.62 feet (800 mm) is covered at outer with galvanized metal sheet having Density 7260 kg/m$^3$, Specific heat 209.34 J/kg-k and Thermal conductivity is taken as 67 w/m-k [Fig. 4] . This galvanized metal sheet is available everywhere in Bastar region. Replacing Ooi's suggested wooden box at 300mm from ground in this solar dryer mesh platform will be used to maintain the air flow inside the solar dryer. The galvanized metal sheet is fitted on a mild steel angle frame and the whole structure can be moved 360° as in bottom of the structure wheels will be provided. At the bottom of the solar dryer from all side 50 mm passage will be provided for air inlet. The 360° movement of solar dryer insures all time maximum sun facing during the solar irradiance. At the lower inclined portion a roof top wind ventilator will be provided to maintain the humidity level inside the box dryer. To control the air flow inside the box dryer a sliding door will be provided just at the bottom of the roof top wind ventilator. A tray will be provided above mesh to check the desired moisture contains texture of the forest produce any point of time. To measure inside temperature and humidity of box solar dryer a battery operated digital temperature gauge and humidity meter can be installed as the product requirement.

*Figure 3.*

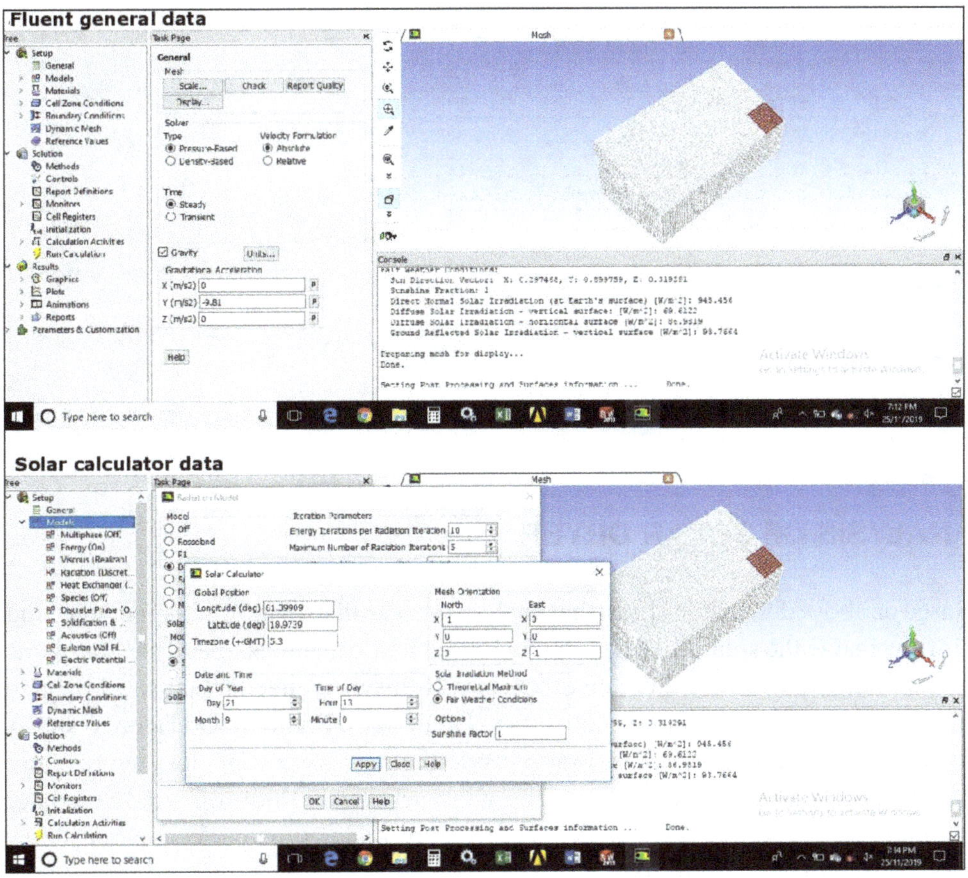

*Figure 4.*

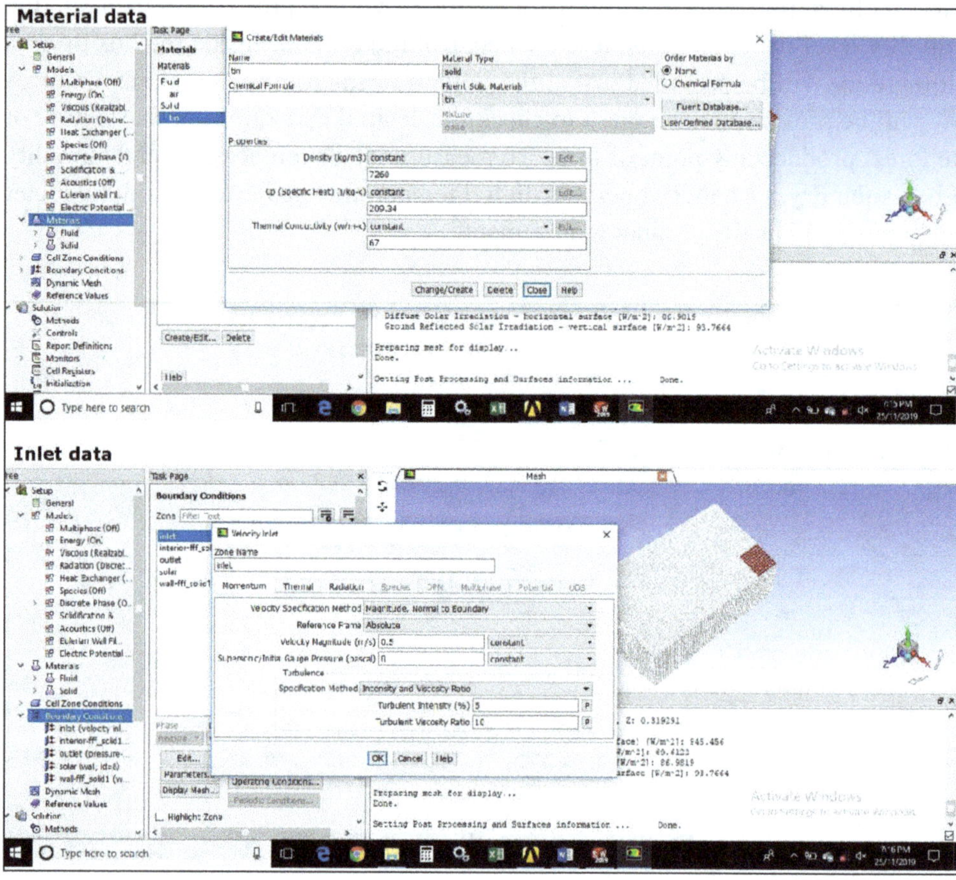


## ANALYSIS OF SOLAR DRYER

Based on above dimension are further analyze for even distribution of temperature and air flow inside the solar dryer. Analysis showed that galvanized metal sheet capable of operating 32 to 45% collection efficiency. If this galvanized metal sheet painted in black color act as a solar heat absorber (Ooi, 2006). Figure 5 and figure 6 shows the evenly distribution of temperature inside the box dryer some less temperature can be visible at the bottom of roof top wind ventilator. Roof top wind ventilator maintains the air velocity inside the dryer. Figure 7 shows the perfect velocity flow of air inside the box solar dryer.

*Figure 5.*

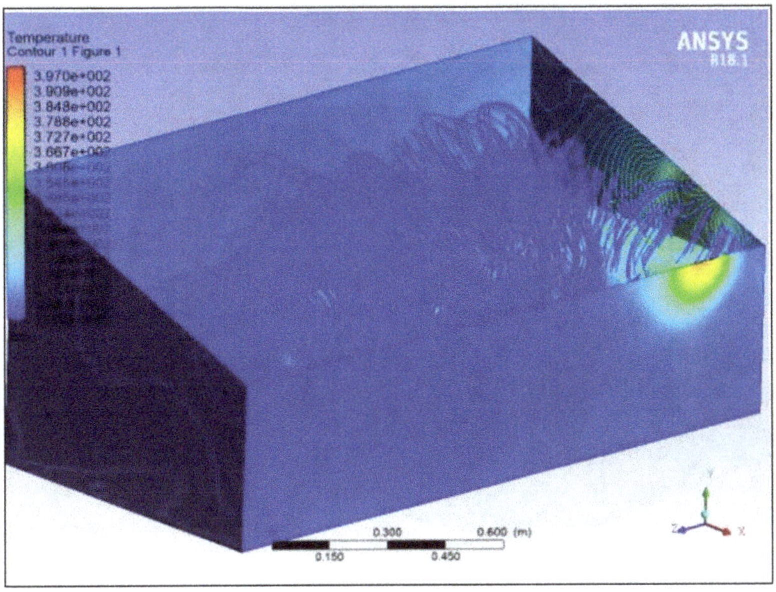


*Figure 6.*

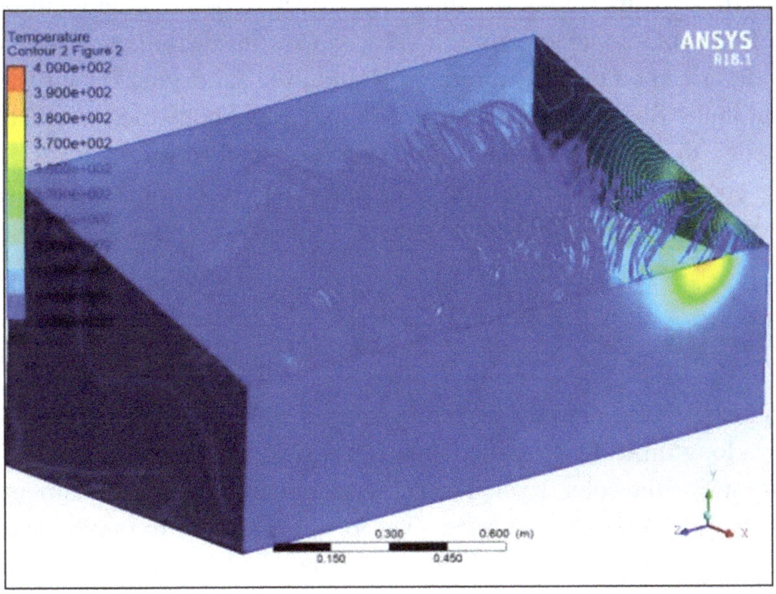

*Figure 7.*

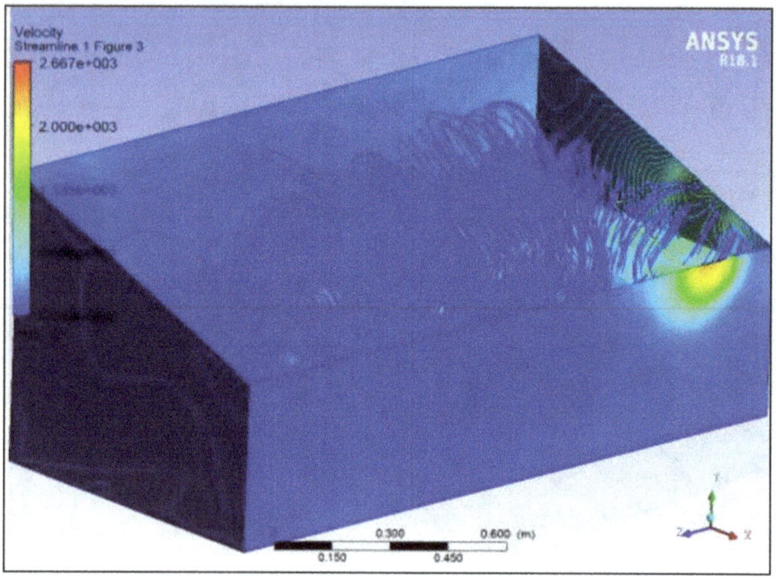


## RESULT ANALYSIS

Above mentioned dimensions are showing perfect computational analysis shown even distribution of temperature and air velocity flow inside the box solar dryer .This box solar dryer can be manufactured easily and used for drying the forest produce throughout in the Bastar region. Simple design, ease of construction and assembly gives an edge to dry the collected forest produce before traded to dry as usual producing open sun drying. The total estimated cost of fabrication and assembly will not be above Rupees 6000/- per dryer.

## CONCLUSION

As the collected forest produce defiantly become valuable and useful if dried and stored for a long time. This ensures the even distributions of temperature and hot air velocity inside the solar dryer. The selected and tested designed dimension also can be easily moved. This solar dryer gives local inhabitants to improve their GMP (Good Manufacturing Practices). That can be manufactured an enhanced to stored valuable dryied collected forest produced to reach at a larger population in future.

## REFERENCES

Aware, R., & Thorat, B. N. (2012). Solar Drying of Fruits and Vegetables. In Solar drying: Fundamentals, Applications and Innovations. Academic Press.

District Forest Office Dantewada. (n.d.). Government of India.

Forest Survey of India. (2011). Government of India.

Hii, C. L., & Law, C. L. (2012). Solar Drying of Major Commodity Products. In Solar drying: Fundamentals, Applications and Innovations. Academic Press.

Ooi, H.S. (2006). Solar Assisted Crop Dryers for Paddy, Flowers and Kenaf. In *SolarAssistedDrying Systems, Innovative TechnologiesforAgricultural andMarine Products*. Islamic Education, Scientific and Cultural Organization.

Sharma, A., Chen, C. R., & Vu Lan, N. (2009). Solar-Energy Drying Systems: A Review. *Renewable & Sustainable Energy Reviews*, *13*(6-7), 1185–1210. doi:10.1016/j.rser.2008.08.015

Tapas, V., & Sharma, P. (2014). The Potentials of Solar Drying in Bastar Region of Chhattisgarh State in India. *International Congress on Renewable Energy (ICORE-2014)*.

Visvale, G. L. (2012). Classification and Selection of Solar Dryers. In Solar drying: Fundamentals, Applications and Innovations. Academic Press.

*Table 1. Processing time and use full parts for forest produce in Bastar regions*

| SN | Botanical Name | Popular Name | Parts Used | Collection Period | Processing Time |
|---|---|---|---|---|---|
| | **Tree** | | | | |
| 1 | Aegle Marmelos | Bel | Fruit/Pulp | March-April | 1 to 3 days |
| 2 | Buchanania Lanzan | Chironjee | Seed | April-May | 5 to 6 days |
| 3 | Cassia Fistula | Amaltas | Pulp | May-July | 2 to 3 days |
| 4 | Phyllanthus Emblica | Amla | Fruit | December-Janurary | 4 to 5 days |
| 5 | Madhuca Indica | Mahua | Flower | February-April | 2 to 3 days |
| | | | seed | May-July | 5 to 6 days |
| 6 | Sapindus Emarginatus | Ritha | Fruit | November- Janurary | 2 to 3 days |
| 7 | Semecarpus Anacardium | Bhilwa | Fruit | Janurary-February | 5 to 6 days |
| 8 | Terminalia Bellerica | Baheda | Fruit | December-Janurary | 2 to 3 days |
| 9 | Butea Monosperma | Palash | Flower | March-April | 1 to 2 days |
| | | | Fruit | May- June | 2 to 3 days |
| 10 | Tamarrindus Indica | Imli | Fruit | March-April | 7 to 12 days |
| | **Shrub** | | | | |
| 1 | Embelia Tsjerium-Cottam | Baibeeding | Fruit | December-Janurary | 1 to 2 days |
| 2 | Helicteres Isora | Marodfalli | Fruit | February-April | 2 to 3 days |
| 3 | Solanum Surattense | kateri | Hole Part (Except root) | February-April | 2 to 3 days |
| 4 | Vitex Negundo | Nirgundi | Leaf | September- November | 1 to 2 days |
| 5 | Woodfordia Fruticosa | Dhawai phool | Flower | February-March | 1 to 2 days |
| | **Herb** | | | | |
| 1 | Acorus Calamus | Bach | Raizom | Janurary-February | 1 to 2 days |
| 2 | Andrographis Paniculata | Kalmegh | Hole Part (Including root) | November- December | 1 to 2 days |
| 3 | Boerhavia Diffusa | Punarnava | Root | May-July | 3 to 4 days |
| 4 | Bacopa Monnieri | Mandukparni | Leaf | August-October | 1 to 2 days |
| 5 | Chlorophytum Borivillianum | Safed Musli | Bulb | October-November | 1 to 2 days |
| 6 | Curculigo Orchioides | Kali Musli | Bulb | February-April | 1 to 2 days |
| 7 | Curcuma Angustifolia | Tikhur | Bulb | November-Janurary | 3 to 4 days |
| 8 | Eclipta Alba | Bhrinngraj | Hole Part (Except root) | July-September | 1 to 2 days |
| 9 | Ocimum Gratissimum | Van Tulsi | Seed | February-April | 2 to 3 days |
| 10 | Phyllanthus Amarus | Bhui Amla | Hole Part (Except root) | September- December | 1 to 2 days |
| 11 | Solanum Nigrum | Makoi | Hole Part (Except root) | February-April | 1 to 2 days |
| 12 | Tribulus Terrestris | Gokshur | Fruit | November-Janurary | 2 to 3 days |
| 13 | Vernonia Anthelmintica | Van Jeera | Seed | November-December | 2 to 3 days |

(Tapas & Sharma, 2014)

# APPENDIX

# Chapter 13
# Characterization of CH3CH2NH3SnI3/ TiO2 Heterojunction:
## Lead-Free Perovskite Solar Cells

**Aloke Verma**
https://orcid.org/0000-0003-4583-5987
*Kalinga University, India*

**A. K. Diwakar**
*Kalinga University, India*

**R. P. Patel**
*Guru Ghasidas University, India*

## ABSTRACT

*In this chapter, CH3CH2NH3SnI3/TiO2 heterojunction lead-free perovskite solar cell (HLFPSCs) is produced by the deposition of perovskite nanoparticles from a solution of CH3CH2NH3I and SnI3 in γ-hydroxyl butyric acid lactone on a 400 nm nanolayers of TiO2 showing (001) planes. The CH3CH2NH3SnI3 /TiO2 HLFSCs shows remarkable PV performance, with Jsc= 17.05 mA/cm2, Voc = 0.71 V, and FF = 0.61; resultant PCE of 6.5% in standard AM 1.5 G one-sun light intensity and PCE of 7.5% in lower light intensity was reported in this work. The beginning of such a simple solution treated HLFPSCs paves the way to appreciate low-cost, high-efficiency LFPSCs.*

DOI: 10.4018/978-1-7998-7611-3.ch013

## INTRODUCTION

Nanocrystalline pigments give the large optical cross-section are harvesting of sunlight in the solar cell conversion system (Goesmann & Feldmann, 2010). Currently, many works are focused on used up device of heterojunction, engineering an intermediate extensive bandgap oxide-based semiconductor such as $TiO_2$ or ZnO as a thin-film layer between the QDs and the oxide based accompanying transparent current collector. Here, we use $TiO_2$ in this work for its large optical behavior. Productivities of 5–6% be situated noted with these unpretentious buildings. In more of a hand in hand QDs solar cell with the unpretentious buildings turned out to be revolved. Also noted of multiple excitation generation (MEG) in QDs based solar cell structure. The corresponding heterojunction QD solar cells show auspicious photovoltaic performance, they still face problems such as low stability, $V_{oc}$, and fast carrier recombination, which prevent them from accomplishing higher efficiencies. These perovskite Nano pigments are remarkably easy to prepare and to put together by unpretentious solution giving out by spin- or dip coating.

## METHOD

The perovskite precipitates from a solution of $SnI_3$ and $CH_3CH_2NH_3I$, forming $CH_3CH_2NH_3SnI_3$ crystals within the concentrates and on the face of it of the $TiO_2$ substrate (Verma et al., 2019). Performance, with $J_{sc}$ of 16.10 mA $cm^{-2}$, a file-factor (FF) of 0.57, and $V_{oc}$ of 0.63 V, conforming PCE of 5.5% under a sun power (intensity) of 1. Earlier reported on the successful use of $CH_3CH_2NH_3SnI_3$ at the same time as an uptake and majority carrier in a HSCs. Here, blend and deposition of $CH_3CH_2NH_3SnI_3/TiO_2$ film were performed by spin coating of a 40-wt% base solution of $CH_3CH_2NH_3I$ and $SnI_3$ in γ- hydroxyl butyric acid lactone (Boopathi et al., 2015). Upon exposure to air at ambient temperature, the film-coated on top of the $TiO_2$ pitch-black in color, showing the materialization of $CH_3CH_2NH_3SnI_3$ in the state of solid, complete by XRD (from Figure 1) (Im et al., 2012).

The sunlight is absorbing by the $CH_3CH_2NH_3SnI_3/TiO_2$ nanoparticles deposited by the spin-coating method. Figure 1 shows a typical XRD pattern of $TiO_2$ nanosheets synthesized at 180 °C. The peaks of $CH_3CH_2NH_3SnI_3$ on the microscope sheet match closely with those of $CH_3CH_2NH_3SnI_3/TiO_2$.

### *Characterization of CH3CH2NH3SnI3/TiO2 Heterojunction*

*Figure 1. XRD pattern of $CH_3CH2NH_3SnI_3/TiO_2$ Heterojunction Lead-Free Perovskite Solar Cell*

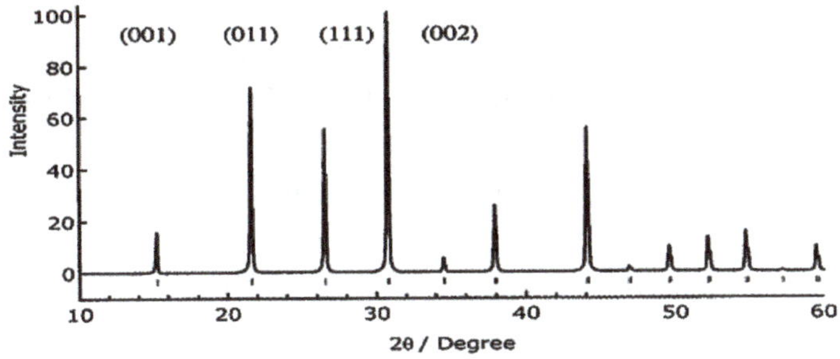


*Figure 2. SEM image of $CH_3CH2NH_3SnI_3/TiO_2$ Heterojunction Lead-Free Perovskite Solar Cell.*

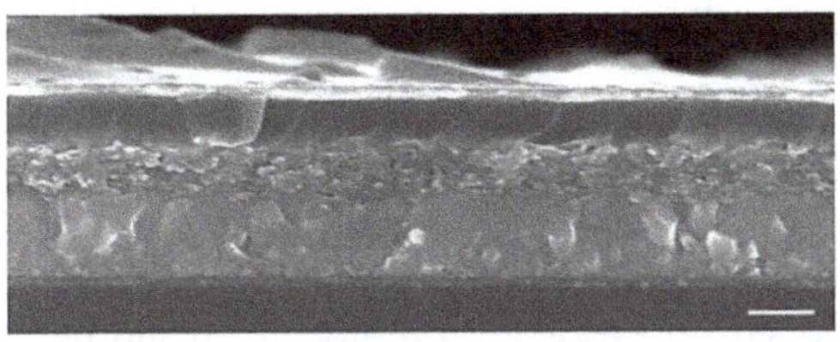

Figure 3A distinguishes itself J–V characteristics of $CH_3CH_2NH_3SnI_3/TiO_2$ heterojunction photovoltaic cell at standard radiance at AM1.5. Here, based on standard reporting conditions, i.e., excellent photocurrent response from 400 to 800 nanometer, Figure 3B shows the IPCE getting a max of 90% in the range of wavelength of 400–540 nanometer and reducing at elongated wavelengths until 800 nanometers. Photocurrent density of 16.2 mA/cm$^2$, in evenhanded arrangement with the result. Succeeded a PCE of only 2.8% at standard illumination at AM1.

*Figure 3. (A) the HLFPSCs J–V characteristic (B) the device spectrum of IPCE.*

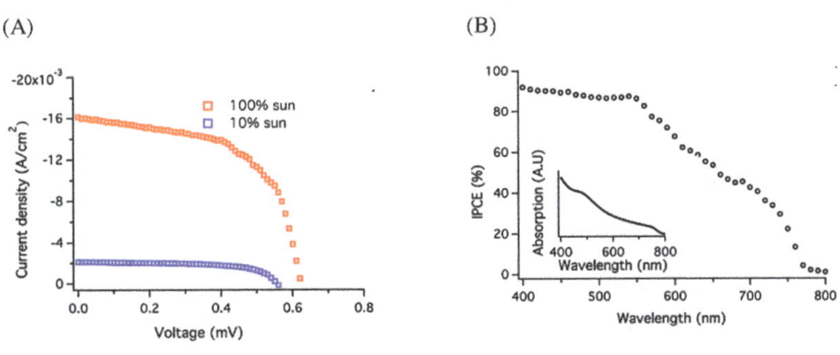

## RESULT AND DISCUSSION

The characteristics of crystallographic of the lead-free-perovskite structures (LFPs) deposited on FTO/TiO2 and a microscopic sheet were examined via XRD and pattern is shown on the Figure 1 (Zhao et al., 2017). The deposition of the $TiO_2$ compact layer was 450 °C. $TiO_2$ nanofilms of ~0.5 μm thickness were spin-coated on top of this substrate and $TiO_2$ layer was annealed at 500 °C for 30 min in air. The synthesis of $CH_3CH_2NH_3SnI_3$ on the $TiO_2$ surface was carried out by dipping a 40 wt% forerunner solution of equimolar $CH_3CH_2NH_3I$ and $SnI_3$ in γ- hydroxyl butyric acid lactone onto the $TiO_2$ film. Using glovebox conditions for the formation of film through the 2000 rpm on 30 s spin coating. At room temperature, during the drying process of $TiO_2$ coated film changed the color and the solid-state indicative the formation of $CH_3CH_2NH_3SnI_3$ (Hao et al., 2014; Ke & Kanatzidis, 2019). The area of active is 0.12 $cm^2$. The generated photocurrent measured and obtained the I-V curves through the applying external biased on the cell (Noel et al., 2014).

## CONCLUSION

In this work, we successfully reported a nanocrystalline based on Heterojunction Lead-Free Solar Cell (HLFPSCs) and crystal structure of the synthesized $CH_3$CH2N$H_3SnI_3$ was determined. The optical behavior of $CH_3$CH2N$H_3SnI_3/TiO_2$ at standard illumination at AM1. Under full sun radiation, the $CH_3$CH2N$H_3SnI_3/TiO_2$ HLFSCs showed conversion efficiency of 2.8% overall.

## REFERENCES

Boopathi, K. M., Ramesh, M., Perumal, P., Huang, Y. C., Tsao, C. S., Chen, Y. F., & Chu, C. W. (2015). Preparation of metal halide perovskite solar cells through a liquid droplet assisted method. *Journal of Materials Chemistry. A, Materials for Energy and Sustainability*, *3*(17), 9257–9263. doi:10.1039/C4TA06392A

Cortecchia, D., Dewi, H. A., Yin, J., Bruno, A., Chen, S., Baikie, T., Boix, P. P., Grätzel, M., Mhaisalkar, S., Soci, C., & Mathews, N. (2016). Lead-free MA2CuCl x Br4–x hybrid perovskites. *Inorganic Chemistry*, *55*(3), 1044–1052. doi:10.1021/acs.inorgchem.5b01896 PMID:26756860

Goesmann, H., & Feldmann, C. (2010). Nanoparticulate functional materials. *Angewandte Chemie International Edition*, *49*(8), 1362–1395. doi:10.1002/anie.200903053 PMID:20108287

Hao, F., Stoumpos, C. C., Cao, D. H., Chang, R. P., & Kanatzidis, M. G. (2014). Lead-free solid-state organic–inorganic halide perovskite solar cells. *Nature Photonics*, *8*(6), 489–494. doi:10.1038/nphoton.2014.82

Im, J. H., Chung, J., Kim, S. J., & Park, N. G. (2012). Synthesis, structure, and photovoltaic property of a nanocrystalline 2H perovskite-type novel sensitizer (CH 3 CH 2 NH 3) PbI 3. *Nanoscale Research Letters*, *7*(1), 353. doi:10.1186/1556-276X-7-353 PMID:22738298

Ke, W., & Kanatzidis, M. G. (2019). Prospects for low-toxicity lead-free perovskite solar cells. *Nature Communications*, *10*(1), 965. doi:10.103841467-019-08918-3 PMID:30814499

Noel, N. K., Stranks, S. D., Abate, A., Wehrenfennig, C., Guarnera, S., Haghighirad, A. A., & Petrozza, A. (2014). Lead-free organic–inorganic tin halide perovskites for photovoltaic applications. *Energy & Environmental Science*, *7*(9), 3061–3068. doi:10.1039/C4EE01076K

Verma, A., Diwakar, A. K., & Patel, R.P. (2019). Synthesis and Characterization of High-Performance Solar Cell. *International Journal of Scientific Research in Physics and Applied Sciences, 7*(2).

Zhao, X. G., Yang, J. H., Fu, Y., Yang, D., Xu, Q., Yu, L., & Zhang, L. (2017). Design of lead-free inorganic halide perovskites for solar cells via cation-transmutation. *Journal of the American Chemical Society*, *139*(7), 2630–2638. doi:10.1021/jacs.6b09645 PMID:28112933

# Chapter 14
# Design of a New Single Phase 7-Level Reduced Switches Count Multilevel Inverter Topology

**Nikhil Agrawal**
*SR Group of Institutions, India*

**Dev Prakash Chourasia**
*Rajiv Gandhi Proudyogiki Vishwavidyalaya, India*

**Bharat Bhushan Khare**
https://orcid.org/0000-0001-8755-9808
*Rajiv Gandhi Proudyogiki Vishwavidyalaya, India*

## ABSTRACT

*This chapter presents a new single-phase multilevel inverter topology. This topology requires fewer components in comparison to conventional multilevel inverter topology such as diode clamped, flying capacitor, and cascade h-bridge inverter. In this topology, no extra clamping diode and no extra clamping capacitor are required for operation. This chapter also presents the control and modulation techniques. These techniques applied on this proposed MLI topology and simulation result are calculated with the help of MATLAB/SIMULINK software and are compared on the basis of THD with over and under modulation index.*

DOI: 10.4018/978-1-7998-7611-3.ch014

## 1. INTRODUCTION

Inverter is a Converter which converts, DC power into AC power at required manner of frequency and magnitude. Multilevel inverter is the extended form of inverter. Multilevel inverter emerged in the market during the 1970-1980s (Agrawal, 2020). Multilevel inverter overcome the problem which is associated with the inverter. The output of inverter is square wave alternating current which have multiple frequencies composition. These multiple frequency components degrades the fundamental component value and create the severe power quality issue. The performance of inverter is determine by its total harmonic distortion in the alternating current side. The inverter output have approx. 48 percent total harmonic distortion which is much higher than the IEEE limit (Agrawal et al., 2017). To replace this inverter, multilevel inverter is used which is start with three level. The multilevel inverter synthesized the square wave and produce the sine wave alternating current. So the output of MLI have lower value of THD. As level increases, the multilevel inverter produce less distorted output waveform but the main problem associated with multilevel inverter is as level increases the components require become more and these components directly affect the cost and complexity. To maintain these two term quality and cost, the optimized topology used with some control techniques to fit between these two term. Conventional multilevel inverter is flying capacitor, diode clamped and cascade H-bridge. In this paper a new multilevel inverter topology proposed which uses less number of switches count in comparisons to conventional MLI topology. In this paper proposed topology shows with some control techniques and determine the simulation result with level shifted PWM techniques and with different modulation index. Modulation index is nothing but the ratio of amplitude of reference signal and amplitude of peak to peak carrier signal. In this paper proposed MLI topology is used for 7 level and shows simulation result and comparisons with conventional MLI topology.

## 2. PROPOSED TOPOLOGY

Proposed topology structure is shown in the figure. The proposed topology have simple configuration of eight switch and three voltage source (Thiyagarajan et al., 2019). Here voltage source have same dc link voltage so proposed topology is the symmetrical MLI topology. The maximum voltage generated across the load is three times the magnitude of voltage source. This is the extendable topology. With this topology we can obtain higher level output just adding the switches to main configuration part. In this topology the witch is IGBT. The switching scheme is shown in table 1. At any level only four switch is ON at a time for voltage level generation.

*Figure 1. Circuit Diagram Proposed Topology*

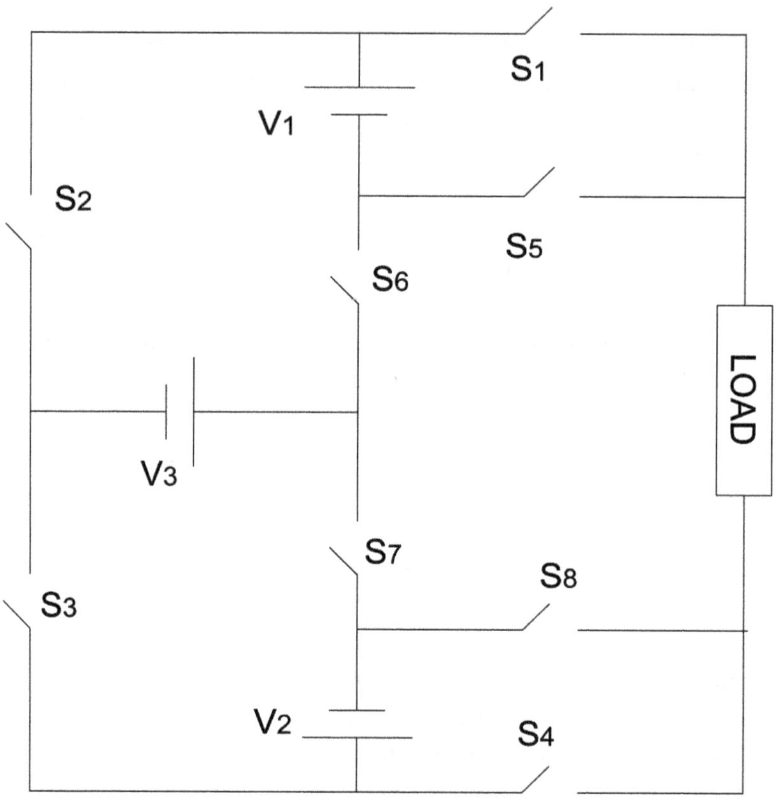


*Table 1. Switching Configuration of Proposed Topology*

| Output Voltage Level | Switches | | | | | | | |
|---|---|---|---|---|---|---|---|---|
| | $S_1$ | $S_2$ | $S_3$ | $S_4$ | $S_5$ | $S_6$ | $S_7$ | $S_8$ |
| $+3V_{dc}$ | ON | OFF | ON | OFF | OFF | ON | OFF | ON |
| $+2V_{dc}$ | ON | OFF | ON | ON | OFF | ON | OFF | OFF |
| $+V_{dc}$ | ON | OFF | OFF | OFF | OFF | ON | ON | ON |
| 0 | ON | OFF | OFF | ON | OFF | ON | ON | OFF |
| $-V_{dc}$ | OFF | ON | ON | ON | ON | OFF | OFF | OFF |
| $-2V_{dc}$ | OFF | ON | OFF | OFF | ON | OFF | ON | ON |
| $-3V_{dc}$ | OFF | ON | OFF | ON | ON | OFF | ON | OFF |

## 3. OPERATING MODES

The operating modes is shown in the figure 2. The darkest line shows the flow of current. Darkest part is the activated part and dotted part is the deactivated part of the circuit.

## 4. CONTROL TECHNIQUES

Pulse Width Modulation Techniques is used for control controlling the output and for improvement of power quality in load side. In this paper level shifted pulse width modulation techniques are used. In the level shifted pulse width modulation techniques K-1 carrier signals are used for K level inverter and one sinusoidal wave as a reference signal. In this paper MLI topology proposed for 7 level so here 6 carrier signals are used.

### 4.1 Phase Disposition (PD PWM)

In Phase Disposition PWM techniques are carrier have same amplitude and arranged with same frequency in same manner (Tomar et al., 2017) as shown in figure 3a and 3b.

### 4.2 Phase Opposition Disposition (POD PWM)

In Phase Opposition Disposition PWM techniques all carrier signals have same amplitude and arranged with same frequency in alternate manner with respect to zero reference line (Najafi & Yatim, 2012) as shown in figure 4a and 4b.

### 4.3 Alternate Phase Opposition Disposition (APOD PWM)

In Alternate Phase Opposition Disposition PWM technique all carrier signals have amplitude and arranged with same frequency in alternate manner with respect to each other (Agrawal et al., 2020) as shown in figure 5a and 5b.

## 5. SIMULATION RESULT

Fig 6 (a) (b) (c) (d) shows Simulation result for Proposed MLI topology.

*Figure 2. (a) (b) (c) (d) (e) (f) (g) are operating Modes of proposed topology*

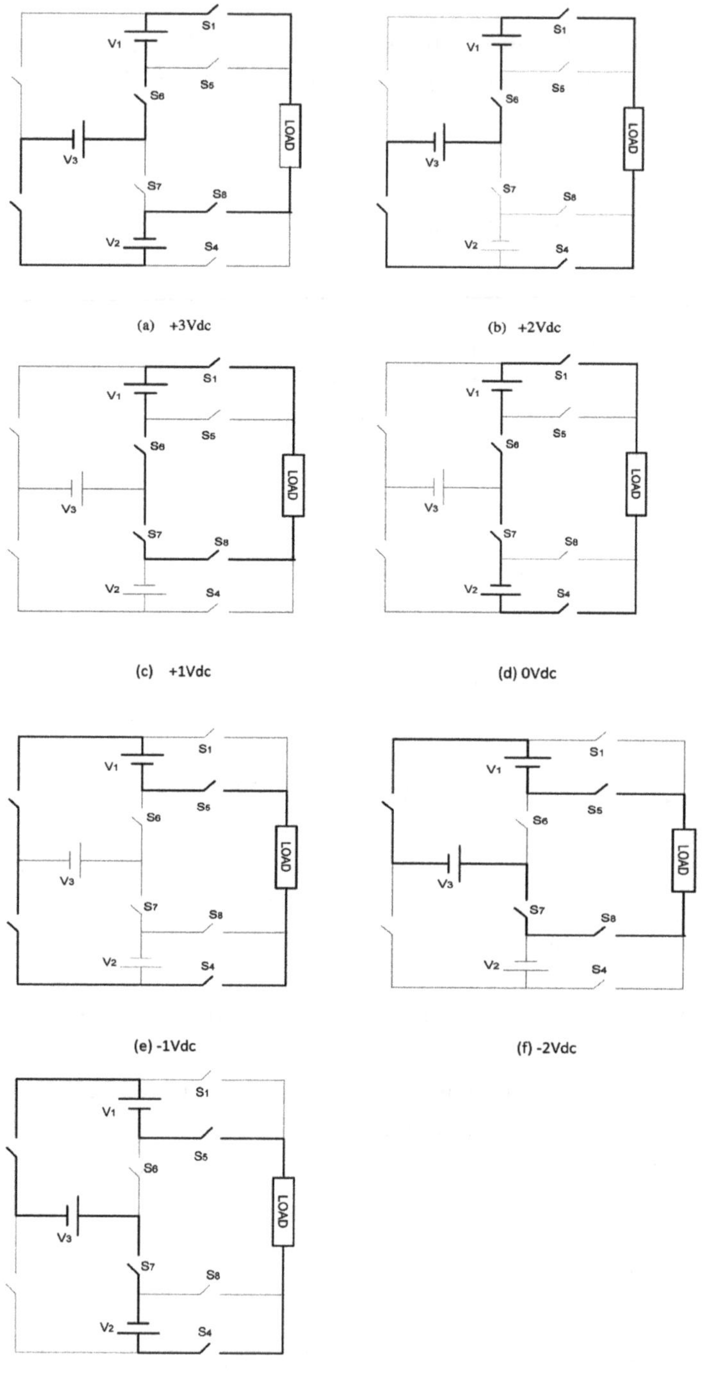


(a) +3Vdc

(b) +2Vdc

(c) +1Vdc

(d) 0Vdc

(e) -1Vdc

(f) -2Vdc

(g) -3Vdc

*Table 2. Components Count in Proposed MLI topology and Conventional MLI topology*

| Components | Diode Clamped | Flying Capacitor | Cascade H-Bridge | Proposed Topology |
|---|---|---|---|---|
| Main Switch | 12 | 12 | 12 | 8 |
| Clamping Diodes | 30 | - | - | - |
| DC Split Capacitor | 6 | 6 | - | - |
| Clamping Capacitor | - | 15 | - | - |
| DC Source | 1 | 1 | 3 | 3 |
| Total | 49 | 34 | 15 | 11 |

*Figure 3a. Carrier Arrangement in PDPWM*

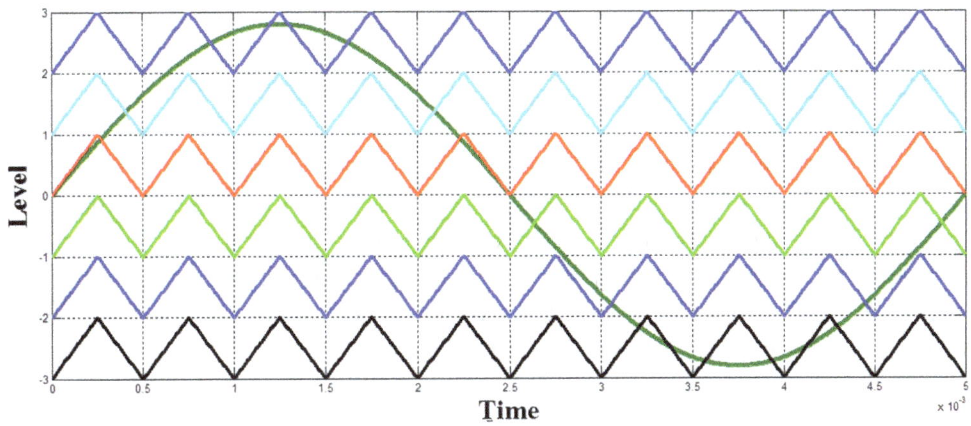


*Figure 3b. Output Voltage in PDPWM*

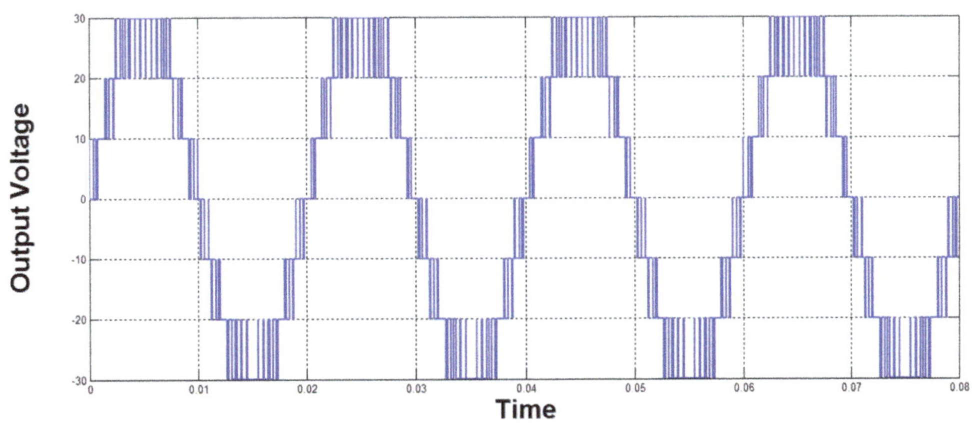

*Figure 4a. Carrier Arrangement in PODPWM*

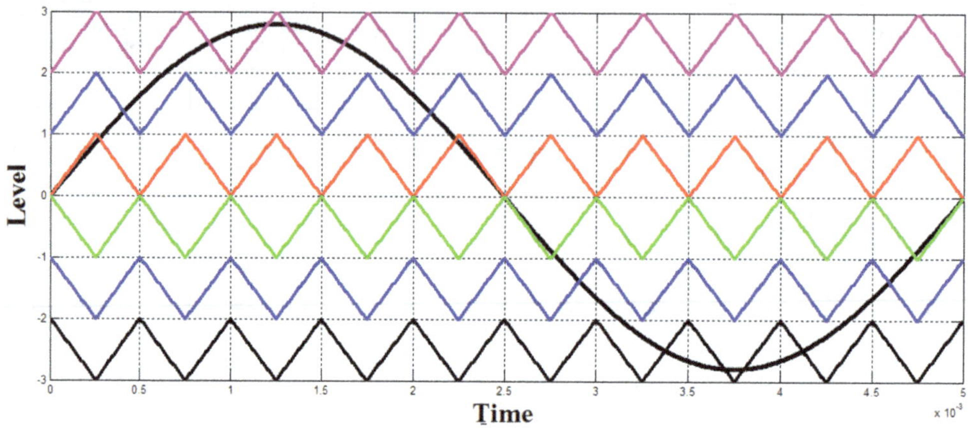


## 6. CONCLUSION

A new MLI topology presented for seven level in this paper. In this paper a comparisons are made between the proposed topology and conventional topologies in symmetric manner. This comparison study shows that proposed MLI topology require less number of switches count. The performance of proposed topology is analyzed by MATLAB/SIMULINK 2010b software. In proposed topology the THD value is obtain is nearly approx. to IEEE standard.

*Figure 4b. Output Voltage in PODPWM*

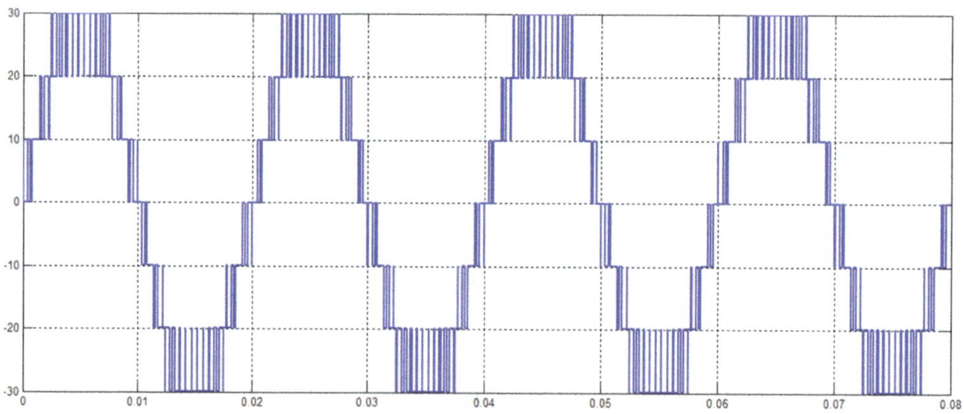

*Figure 5a. Carrier Arrangement in APODPWM*

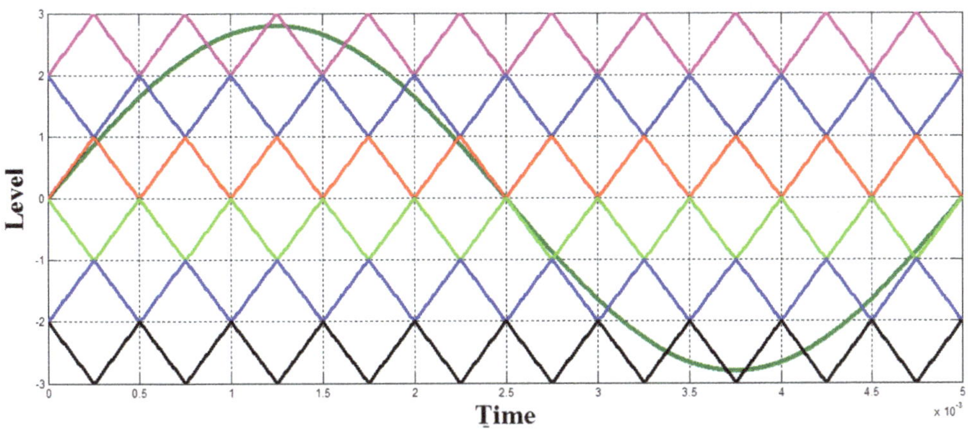


*Figure 5b. Output Voltage in APODPWM*

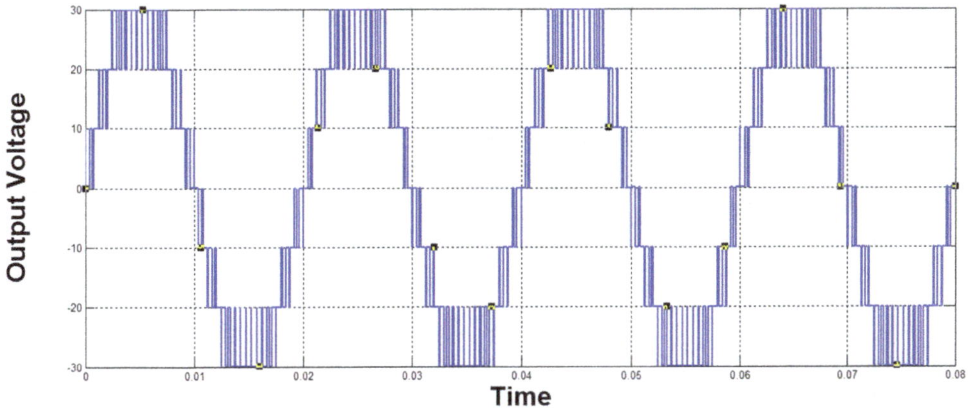


## ACKNOWLEDGMENT

The author thankful to Electrical Engineering Department & Research and Development Department, S R Group of Institutions Jhansi, India to the guidance and support for carrying out this work.

*Figure 6a. FFT of Proposed Topology with PDPWM and Ma=1.1*

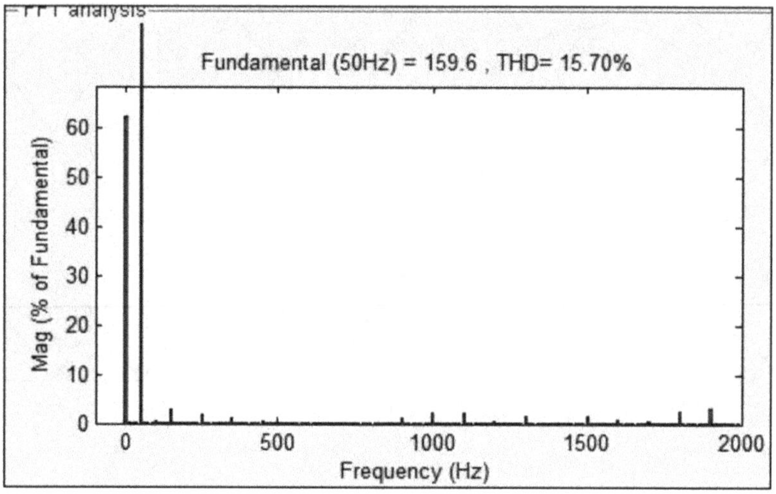


*Figure 6b. FFT of Proposed Topology with PDPWM and Ma=1.0*

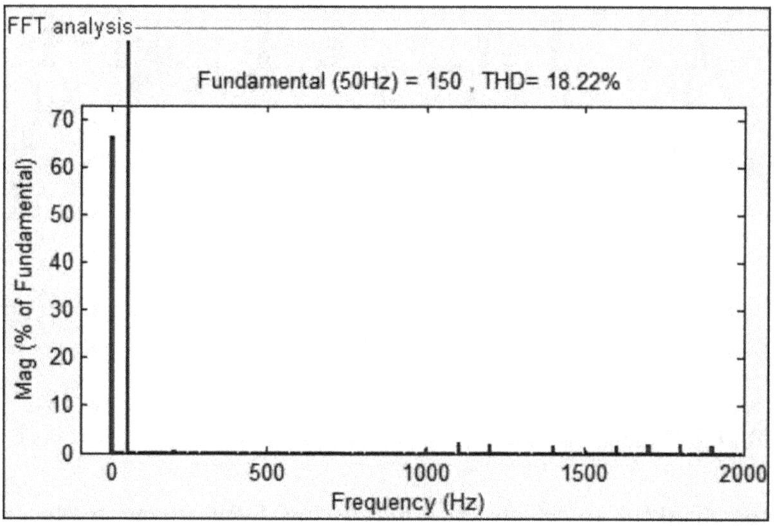

*Figure 6c. FFT of Proposed Topology with PODPWM and Ma=1.1*

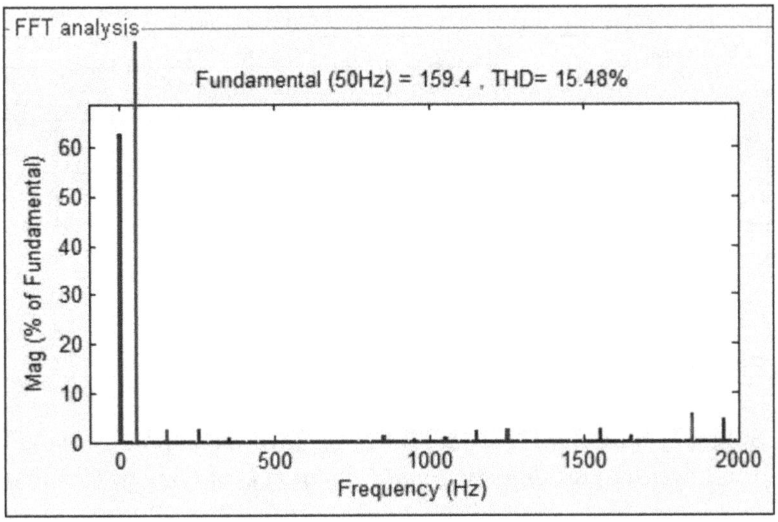


*Figure 6d. FFT of Proposed Topology with APODPWM and Ma=1.1*

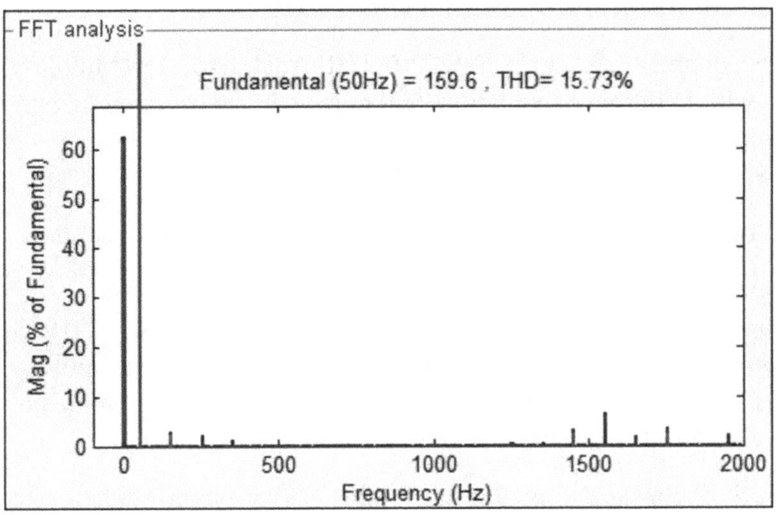

*Table 3. Proposed Topology result for 7 level inverter*

| Proposed Topology Level | Modulation Index | PWM Techniques | | | | | |
|---|---|---|---|---|---|---|---|
| | | PD | | POD | | APOD | |
| | | % THD | Fundamental Component | % THD | Fundamental Component | % THD | Fundamental Component |
| 7 Level | 1.1 (Over Modulation) | 15.70 | 159.6 | 15.48 | 159.4 | 15.75 | 159.6 |
| | 1.0 (Unity Modulation) | 18.22 | 150 | 18.13 | 149.8 | 18.34 | 149.9 |
| | 0.9 (Under Modulation) | 22.38 | 135.2 | 22.06 | 135.6 | 21.93 | 135.6 |

## REFERENCES


Agrawal, Singh, & Bansal. (2017). A Multilevel Inverter Topology using Reverse-Connected Voltage Sources. *International Conference on Energy, Communication, Data Analytics and soft Computing (ICECDS 2017),* 1290-1295.

Agrawal, N. (2020). *Performance Analysis of Cascade H-Bridge Multilevel Inverter Topology With Filter Circuit and Without Filter Circuit. In Design and Optimization of Sensors and Antennas for Wearable Devices: Emerging Research and Opportunities.* IGI Global. doi:10.4018/978-1-5225-9683-7.ch008

Agrawal, N., Bansal, P., & Umale, N. (2020). THD Analysis of New Multilevel Inverter Topology with Different Modulation Techniques. In *International Conference on Intelligent Computing and Smart Communication 2019. Algorithms for Intelligent Systems*. Springer. 10.1007/978-981-15-0633-8_117

Najafi & Yatim. (2012). Design and Implementation of a new multilevel inverter topology. *IEEE Transactions on Industrial Electronics, 59*(11), 4148-4154.

Thiyagarajan, V., Somasundaram, P., & Ramash Kumar, K. (2019, June 15). Simulation and Analysis of Novel Extendable Multilevel Inverter Topology. *Journal of Circuits, Systems, and Computers*, *28*(6), 1950089. Advance online publication. doi:10.1142/S0218126619500890

Tomar, S. S., Agrawal, N., & Bansal, P. (2017). Various Modulation techniques in symmetrical multilevel inverters. *International Conference on Energy, Communication, Data Analytics and soft Computing (ICECDS 2017),* 1322-1327.

# Chapter 15
# Racket Shape Textile Antenna for Triple-Band Frequency Applications

**Akanksha Lohia**
*SR Group of Institutions, India*

**Jyotsna Pal**
*SR Group of Institutions, India*

**Nishi Kushwaha**
*SR Group of Institutions, India*

**Supriya Shreevas**
*SR Group of Institutions, India*

## ABSTRACT

*The chapter explores a racket shaped antenna that operates in triple band frequency configuration. The main advantage of this antenna is that the design uses wearable textile material as a substrate. Here, jeans material is used as a substrate whose dielectric constant is 1.7 and thickness is 1mm. The designing and simulation are done with the help of CST software. The anticipated antenna operates in C-band and X-band providing three resonant frequencies 7.335GHz, 8 GHz, and 8.38 GHz. Different analyses have been done using CST software based on parameters like return loss, radiation pattern, smith chart, polarization, etc. For obtaining the best result, parametric study has also been done by varying the geometry of the design.*

DOI: 10.4018/978-1-7998-7611-3.ch015

## 1. INTRODUCTION

An antenna is a transceiver (transmitter + receiver) device. Antenna is used for many application such as biomedical science (to cure breast cancer is the latest topic), Satellite communication, Defense radar and remote sensing according to frequency range that it is operating (Singh et al., 2015a; Singh et al., 2015b; Singh et al., 2017; Singh et al., 2020; Yadav, Kumar Singh, Kumar Bhoi et al, 2020; Yadav & Singh, 2019; Yadav, Singh, Yadav et al, 2020) .Wireless transmission has become our preference because of its several advantages mainly high efficiency due to reduction in losses, low maintenance and cost effective. Now days, a very interesting phenomenon plays an important role in the industry that is the use of textile material. Textile material turns antenna to operate with human body so smoothly. It makes antenna wearable, light in weight and provide flexibility(Kalpana, 2017; Khare et al., 2019; Lohia et al., 2020; Singh, Ali, Singh et al, 2013; Singh, Saxena, Khare, Shayka et al, 2018; Singh, Kabeer, Shukla et al, 2013; Singh et al., 2014a; Singh & Saxena, 2018; Yadav et al., 2019).Because of these reasons, textile antenna has became very popular into our real life such as healthcare applications, military communications and in some risky applications like mining etc. Wearable antenna is designed with the help of clothing. In spite of these advantages, we must be careful in designing the antenna with textile material so that antenna performances should not be affected by bending and shrinking of clothing. Also it should not have negative impact on human body. For this, antenna testing should be done in every typical circumstance (Kalam, Das, Sharma, Kushwaha et al, 2018; Lumba et al., 2014; Saxena, 2019; Saxena & Khare, 2019; Kalam, Das, Sharma, Sharma et al, 2018; Kalam, Das, Sharma, Singh et al, 2018; Singh, Singh, & Khanna, 2018; Singh et al., 2014b; Srivastava et al., 2015).

Here we design a badminton shaped antenna that is small in size. It has used jeans as a substrate material and copper tape for patch and ground plane. It has used microstrip linefeed for input that takes less power and has less impact of radiation on human body. It operates in the frequency range of C-band and X-band applications like satellite communication and weather monitoring etc. Primarily C- band is used for satellite communications for full time TV networks etc. Since it is less susceptible to rainfall so it is used in that area probably (Bano et al., 2018; Dubey et al., 2018; Lala et al., 2018; Saini et al., 2018; Singh, Saxena, Khare, Shakya et al, 2018; Singh, Singh, Kumar et al, 2018; Singh, Singh, Saxena et al, 2018).

The next sections of this paper are as follows:

Section 2, defines antenna structure with its dimensions and also with its evotionary designs, section 3, defines the parametric study of the proposed antenna design by varying its geometry with its optimization results, Section 4, describes the result outcomes of anticipated antenna with different parameters and at last in Section 5, the result of proposed antenna design performances was concluded.

## 2. PROPOSED ANTENNA DESIGN

The proposed antenna design was carried out with the help of iteration steps. Firstly, we have designed the design1 then design2 and then finally design3(proposed design). The design layout has also shown by figure1 (a), (b),(c) and (d).The figure1 (c) and (d) represent the front and back view of proposed design. These designs are made on wearable textile material jeans as a substrate which has dielectric constant 1.7 and thickness 1mm with a partial ground structure. For ground plane and patch, copper tape has been used as a conducting material with thickness 0.038mm.The badminton shape patch is designed with microstrip line feed for the input. The different parameter dimension of the proposed antenna is given in table 1.

*Table 1. Dimension for proposed textile antenna design*

| S.No. | Proposed antenna design parameter | Dimension |
|---|---|---|
| 1 | Ground dimension | 40mm x 14mm |
| 2 | substrate dimension | 40mm x 40mm |
| 3 | Substrate height | 1mm |
| 4 | Patch outer circle radius | 10mm |
| 5 | Radius of outer circle cut in patch | 8mm |
| 6 | Microstrip feed line width | 2mm |
| 7 | Patch strip slot width | 0.8mm |

## 3. PARAMETRIC ANALYSIS OF THE PROPOSED ANTENNA

For this analysis, the different parameters of proposed antennas design are taken. By varying the dimension of these parameters, the optimized design of the proposed antenna is obtained.

### (i) By Varying the Height of Substrate Height (h)

By varying the height of substrate textile material i.e. Jeans, the results are obtained at different resonant frequencies. Figure 2 represents the optimized results through variation in substrate height. At substrate height (h) 0.8mm, 1.2mm and 1mm, the maximum directivity of the designed antenna is 4.538dBi, 4.598dBi and 4.593dBi respectively at different resonant frequencies 7.37GHz, 7.95 GHz and 8 GHz. Since there is a little difference in directivity at 1mm and 1.2mm but there is a significant change in bandwidth. So substrate height of 1mm is selected for proposed antenna design.

*Figure 1. Layout designs of textile antenna (a) design1 (b) design2 (c) Front view of proposed antenna (d) Back view of proposed antenna*

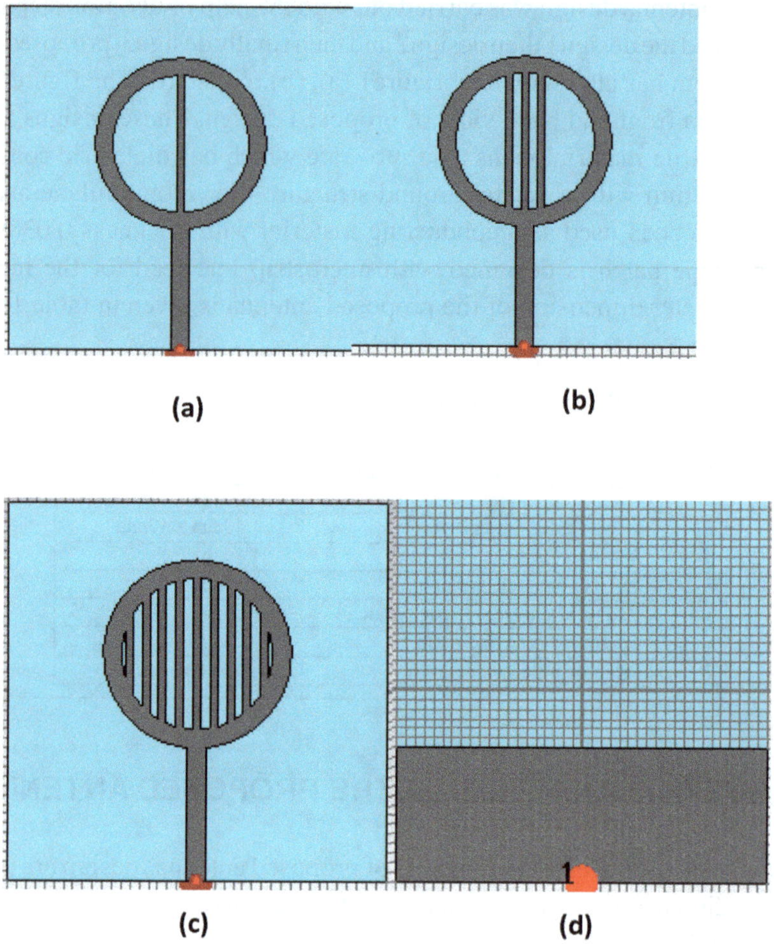


## (ii) By Varying the Width of Microstrip Line Feed (a)

Microstrip line feed is fabricated by using copper tape. By changing its width, the optimized design can be achieved. The results are analyzed at different widths i.e. 1mm, 1.6mm and 2mm at different resonant frequencies. Figure 3 represents the optimized results through variation in the width of line feed width. At line feed width 1mm, 1.6mm and 2mm, the maximum directivity of the designed antenna is 4.548dBi, 4.605dBi and 4.593dBi respectively at different resonant frequencies 7.335GHz, 8 GHz and 8 GHz. So it is clear that line feed width of 2mm is selected for proposed antenna design.

*Figure 2. Optimized Simulation graph of return loss (dB) Vs frequency (GHz) by varying substrate height*

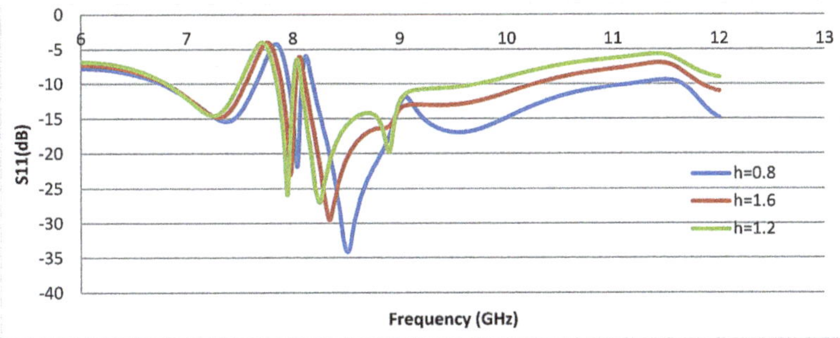


*Figure 3. Optimized Simulation graph of return loss (dB) Vs. frequency (GHz) by varying strip line feed width*

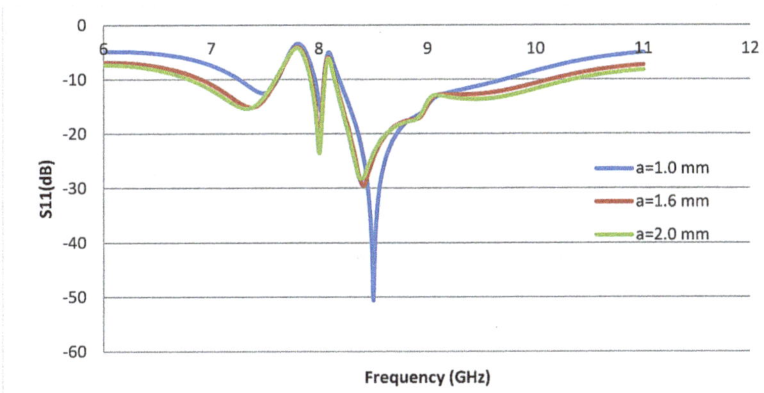


## 4. RESULT ANALYSIS

After designing, the simulation is done using CST software. The results are analyzed and studied based on different antenna performances parameters namely: Return loss, radiation efficiency, radiation pattern, polarization and smith chart.

- **Return loss-**Figure 4 shows the graph between return loss magnitude (dB) vs. frequency (GHz) with multiband configurations at different resonant frequencies 7.335 GHz, 8 GHz and 8.38 GHz respectively of the proposed antenna design.

*Figure 4. Simulation graph of return loss (dB) vs. frequency (GHz)*

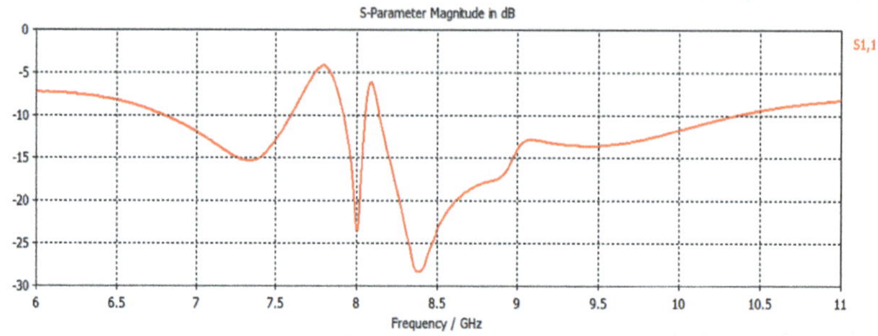


The optimized result of return loss (dB) vs. frequency (GHz) between design1,design 2 and design 3 (proposed design) has been analyzed that are given below in figure 5..In all the three designs, the return loss, directivity and bandwidth is best in the proposed design.

*Figure 5. Optimized Simulation graph of return loss (dB) vs. frequency (GHz) for different designs*

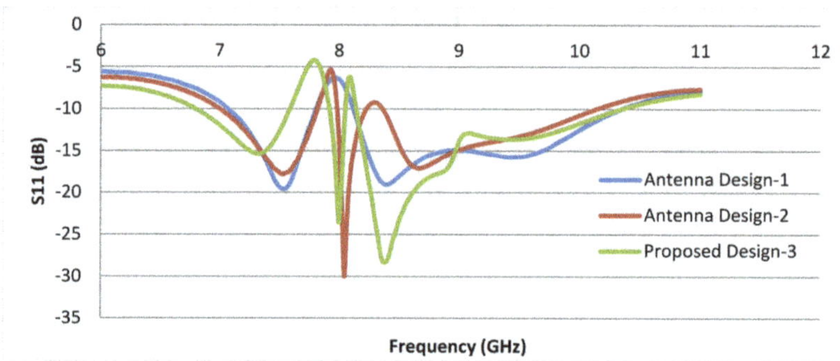


- **Radiation Pattern-**The radiation pattern in 3-D pattern are also analyzed at resonant frequencies 7.335GHz, 8GHZ and 8.38GHz of the proposed antenna design. Figure 5 (a),(b) and (c) represents the radiation pattern and radiation efficiency of the resonant frequencies.

*Figure 6. 3-D radiation pattern at resonant frequencies (a) 7.335GHz (b) 8GHz (c) 8.38GHz*

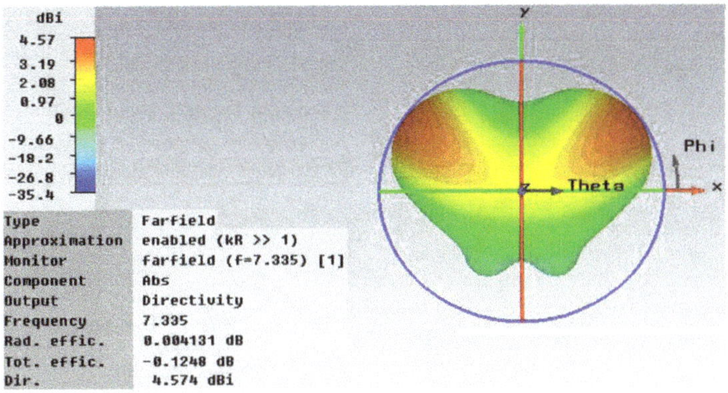


(a)

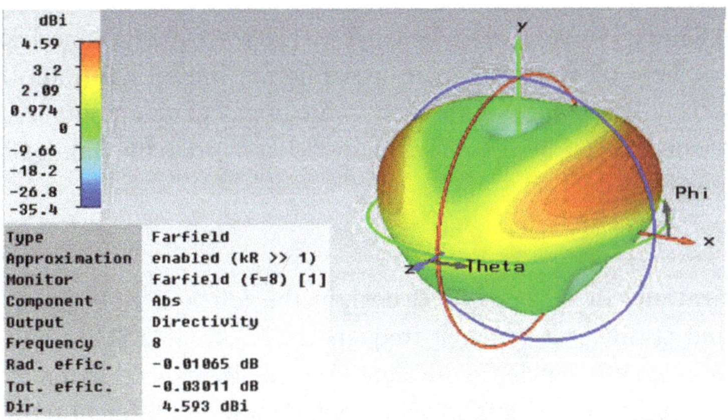


(b)

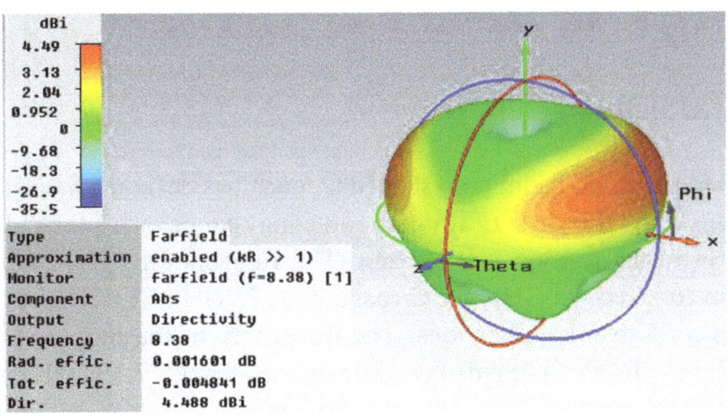


(c)

*Figure 7. Smith Chart of the Proposed antenna*

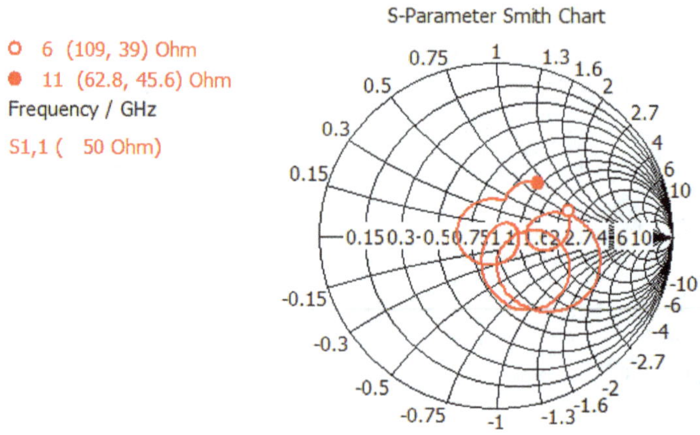


- **Smith Chart-** As we have studied that we can solve many numerical problems with the help of Smith Chart in transmission line in RF (radio frequency) engineering. By using Smith chart, we can calculate admittance, reflection coefficients etc. So, here we have analyzed in figure 6 the Smith Chart for the proposed design.

- **Polarization-** In this proposed design, the left and right polarization are analyzed at different resonant frequencies 7.335 GHz,8GHz and 8.38GHz. The radiation efficiencies at different frequencies and at different polarization are also analyzed. The left and right polarization of different frequencies are presented in figure 7,8,9.

## 5. CONCLUSION

The presented proposed badminton shaped antenna has designed by using a textile material Jeans with a thickness of 1mm.The antenna operates at triple band frequency configuration in the C-band and X-band. The frequency range of the designed antenna is in (6.8GHz-7.6GHz) for C- band and (7.9GHz - 8.05GHz and 8.1GHz -10.35GHz) for X-band applications. The directivity of the proposed antenna is 4.574dBi, 4.593 dBi and 4.488 dBi at resonant frequencies 7.335 GHz, 8GHz and 8.38GHz respectively.

*Figure 8. Polarization at 7.335 GHz (a) Left polarization (b) Right Polarization*

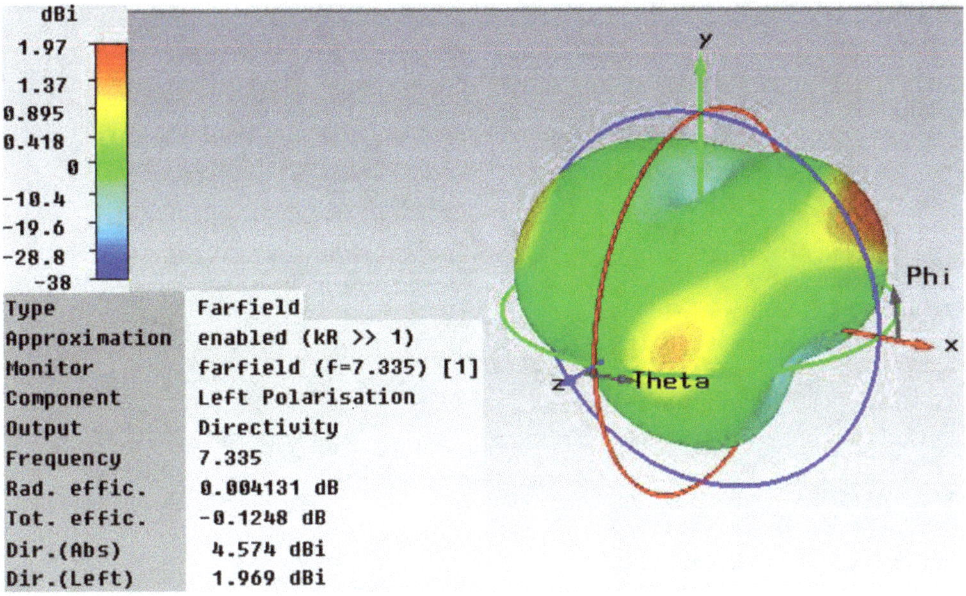


(a)

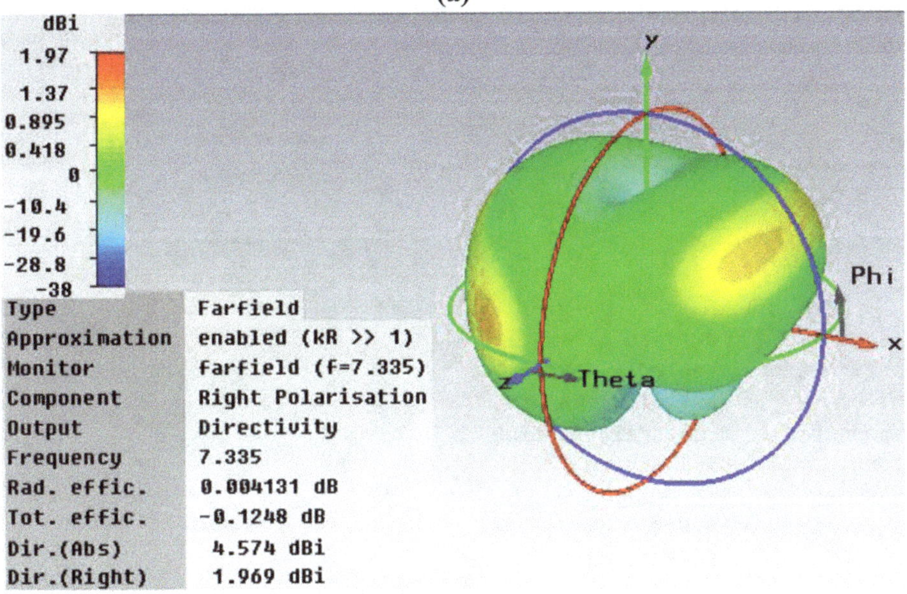


(b)

*Figure 9. Polarization at 8 GHz (a) Left polarization (b) Right Polarization*

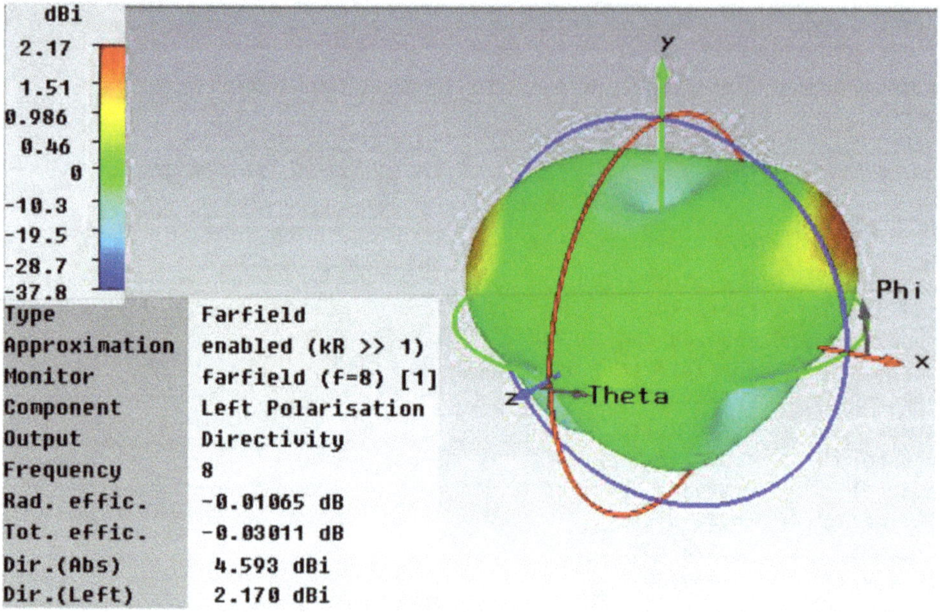


(a)

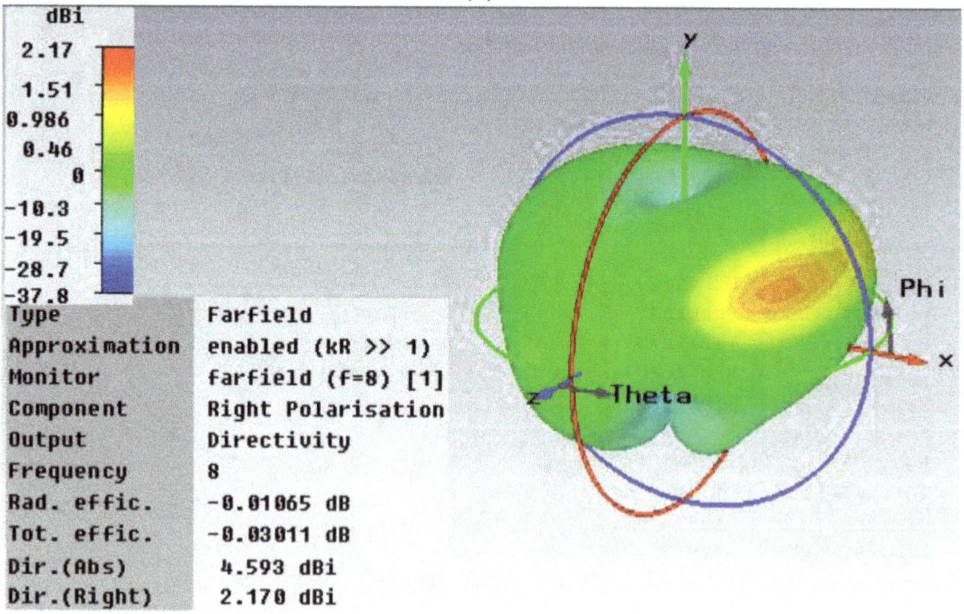


(b)

*Figure 10. Polarization at 8.38GHz (a) Left polarization (b) Right Polarization*

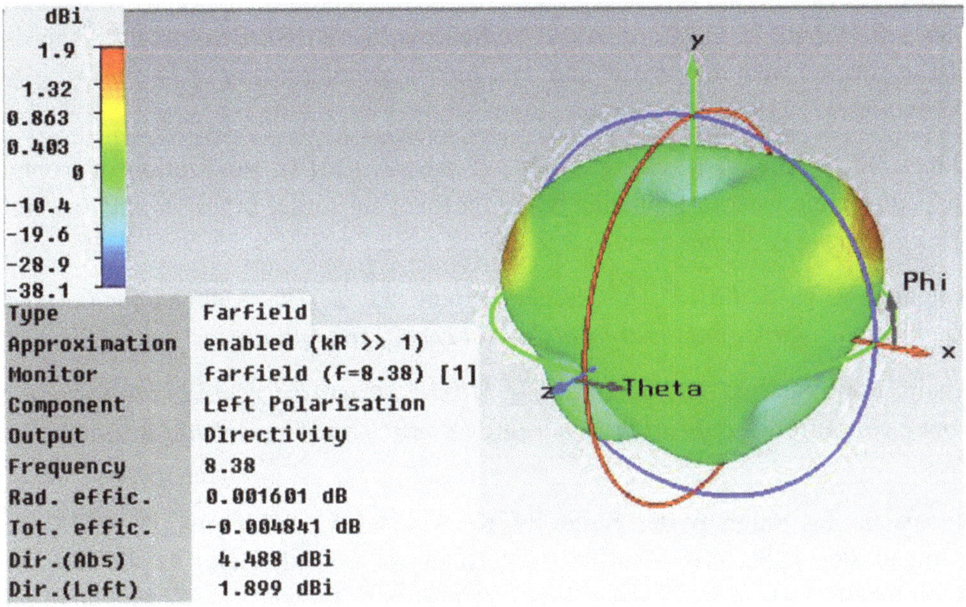


**(a)**

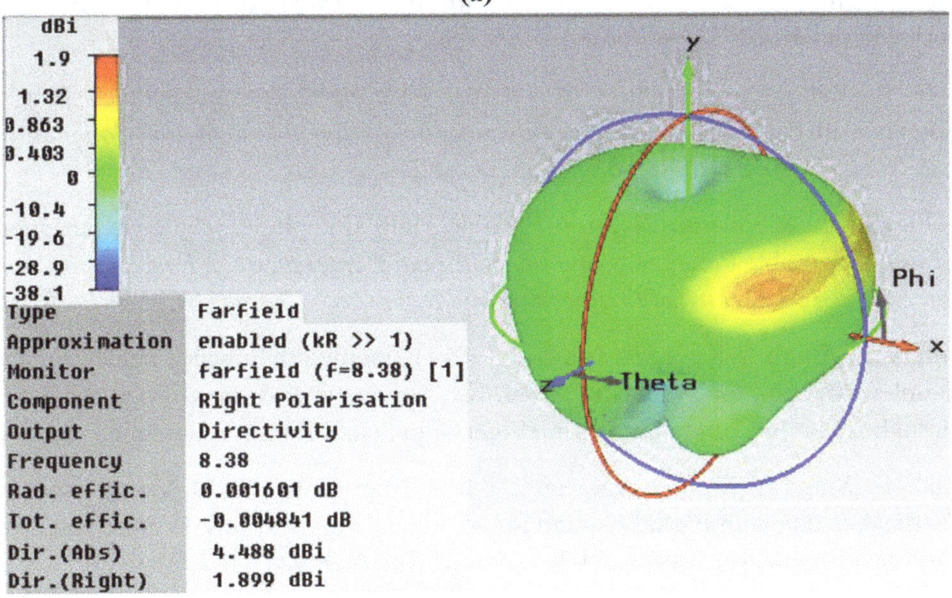


**(b)**

## REFERENCES

Bano, N., Singh, V. K., Bhoi, A. K., & Sharma, S. (2018). Design and Analysis of Key shaped wearable textile antenna. *International Journal of Engineering & Technology, 7*, 219-221.

Dubey, R., Singh, V. K., Bhoi, A. K., & Ali, Z. (2018). Realization of trophy shaped flexible wearable antenna based on foam substrate. *International Journal of Engineering & Technology, 7*, 222-224.

Kalpana, V. K. S. (2017). A Crescent Moon like Textile Antenna for C-Band Application. *International Journal of Control Theory and Application, 10*(9), 899-903.

Khare, Bhoi, Sharma, & Lohia. (2019). *Missile Structured Wearable Antenna for Power Harvesting Application: Emerging Research and Opportunities*. Academic Press.

Kushwaha, R., Singh, V. K., Singh, N. K., Saxena, A., & Sharma, D. (2018). A Compact Pentagonal Textile Microstrip Antenna for Wide Band Application. In A. Kalam, S. Das, & K. Sharma (Eds.), *Advances in Electronics, Communication and Computing. Lecture Notes in Electrical Engineering* (Vol. 443, pp. 793–799). Springer. doi:10.1007/978-981-10-4765-7_82

Lala, K., Lala, A., & Singh, V. K. (2018). Wide Band Triangular Patch Textile Antenna with Partial Ground Plane. *Advances in Intelligent Systems and Computing, 632*, 577-583. 10.1007/978-981-10-5520-1_52

Lohia, Singh, & Saxena. (2020). A Flexible Multi slot wireless Antenna designed for Reconfigurable Frequency. *2nd International Conferences on Emerging Trends and Advances in Electrical Engineering and Renewable Energy (ETAEERE-2020).*

Lumba, S., Singh, V. K., & Srivastava, R. (2014). Bandwidth Enhancement by Direct Coupled Antenna for WLAN/GPS/WiMax Applications & Feed Point Analysis through ANN. In Computational Intelligence in Data Mining. Academic Press.

Saini, R., Singh, V. K., Singh, N., Saini, J. P., & Bhoi, A. K. (2018). Multi Resonant Textile Antenna with Partial Ground for Multiband Applications. Lecture Notes in Electrical Engineering, 462, 359-367. doi:10.1007/978-981-10-7901-6_39

Saxena. (2019). A La Shape Antenna for High Frequencies Application. In *Design and Optimization of Sensors and Antennas for Wearable Devices: Emerging Research and Opportunities*. Academic Press.

Saxena & Khare. (2019). A Hammer type Textile Antenna with partial circle ground for wide band applications. In *Design and Optimization of Sensors and Antennas for Wearable Devices: Emerging Research and Opportunities*. Academic Press.

Sharma, P., Yadav, A., & Singh, V. K. (2018). Design of Circularly Polarized Antenna with Different Iterations for UWB Applications. In A. Kalam, S. Das, & K. Sharma (Eds.), *Advances in Electronics, Communication and Computing. Lecture Notes in Electrical Engineering* (Vol. 443, pp. 441–447). Springer. doi:10.1007/978-981-10-4765-7_47

Singh, Ali, Singh, & Ayub. (2013). Dual Band Triangular Slotted Stacked Microstrip Antenna for Wireless Applications. *Open Engineering, 3*(2), 221-225.

Singh, Saxena, Khare, Shakya, Sharma, & Bhoi. (2018). Power harvesting through flexible rectenna at dual resonant frequency for low power devices. *International Journal of Engineering & Technology, 7,* 647-649.

Singh, Saxena, Khare, Shayka, Chae, Sharma, & Bhoi. (2018). Power Harvesting through flexible rectenna at dual resonant frequency for low power devices. *International Journal of Engineering and Technology, 7*(2.33), 647-649.

Singh, Singh, Kumar, Yadav, & Bhoi. (2018). Rectenna design for electromagnetic energy harvesting and wireless power transfer. *International Journal of Engineering & Technology, 7*, 632-633.

Singh, A. K., Kabeer, R. A., Shukla, M., Singh, V. K., & Ali, Z. (2013). Performance Analysis of First Iteration Koch Curve Fractal Log Periodic Antenna of Varying Angles. *Open Engineering, 3*(1), 51-57.

Singh, N., Singh, A. K., & Singh, V. K. (2015a). Design & Performance of Wearable Ultra Wide Band Textile Antenna for Medical Applications. *Microwave and Optical Technology Letters, 57*(7), 1553-1557.

Singh, N., Singh, A. K., & Singh, V. K. (2015b). Ultra Wide Band Textile Antenna for Medical Applications. *Open Engineering, 5*(1), 117-119.

Singh, N. K., Sharma, N., Ali, Z., Singh, V. K., & Bhoi, A. K. (2018). Inset Fed Circular Microstrip Antenna with Defected Ground. In A. Kalam, S. Das, & K. Sharma (Eds.), *Advances in Electronics, Communication and Computing. Lecture Notes in Electrical Engineering* (Vol. 443, pp. 605–611). Springer. doi:10.1007/978-981-10-4765-7_63

Singh, N. K., Singh, V. K., Saxena, A., Bhoi, A. K., Garg, A., & Sherpa, K. S. (2018). A Compact Slotted Textile Patch Antenna for Ultra-wide Band Application. Lecture Notes in Electrical Engineering, 443, 53-59. doi:10.1007/978-981-10-4765-7_6

Singh, R., Singh, V. K., & Khanna, P. (2018). A Compact CPW-Fed Defected Ground Microstrip Antenna for Ku Band Application. Lecture Notes in Electrical Engineering, 443, 231-237. doi:10.1007/978-981-10-4765-7_24

Singh, V. K., Ali, Z., Ayub, S., & Singh, A. K. (2014a). Bandwidth Optimization of Compact Microstrip Antenna for PCS/DCS/Bluetooth Application. *Open Engineering, 4*(3), 281-286.

Singh, V. K., Ali, Z., Ayub, S., & Singh, A. K. (2014b). A wide band Compact Microstrip Antenna for GPS/DCS/PCS/WLAN Applications. In Intelligent Computing, Networking, and Informatics. Academic Press.

Singh, V. K., Dhupkariya, S., & Bangari, N. (2017). Wearable Ultra Wide Dual Band Flexible Textile Antenna for WiMax/WLAN Application. *International Journal of Wireless Personal Communications, 95*(2), 1075–1086.

Singh, V. K., Lala, A., & Singh, A. K. (2020). Novel Inset Feed Circular Slotted Microstrip Antenna Using Multilayer Feed-Forward Back-Propagation and Radial Basis Function Neural Network. *National Academy Science Letters.*

Singh, V. K., & Saxena, A. (2018). Two Parabolic Shape Microstrip Patch Antenna for Single Band Application. *Journal of Microwave & Technologies, 5*(1), 33–36.

Srivastava, R., Singh, V. K., & Ayub, S. (2015). Comparative Analysis and Bandwidth Enhancement with Direct Coupled C Slotted Microstrip Antenna for Dual Wide Band Applications. In Frontiers of Intelligent Computing: Theory and Applications. Springer.

Yadav & Singh. (2019). Design of U-shape with DGS circularly polarized wearable antenna on fabric substrate for WLAN and C-Band applications. *Journal of Computational Electronics, 18*(3), 1103–1109.

Yadav, Singh, & Mohan. (2019). *Design of a U-shaped Circularly wearable antenna with DGS on a fabric substrate for WLAN and C-band Applications*. Academic Press.

Yadav, A., Kumar Singh, V., Kumar Bhoi, A., Marques, G., Garcia-Zapirain, B., & de la Torre Díez, I. (2020). Wireless Body Area Networks: UWB Wearable Textile Antenna for Telemedicine and Mobile Health Systems. *Micromachines, 11*(6), E558.

Yadav, A., Singh, V.K., Yadav, P., Beliya, A.K., Bhoi, A.K., & Barsocchi, P. (2020). Design of Circularly Polarized Triple-Band Wearable Textile Antenna with Safe Low SAR for Human Health. *Electronics*, *9*(1366).

# Compilation of References


Bushyager, N. A., & Tentzeris, M. M. (2001, May). Modeling and design of RF MEMS structures using computationally efficient numerical techniques. In *2001 Proceedings. 51st Electronic Components and Technology Conference (Cat. No. 01CH37220)* (pp. 327-330). IEEE. 10.1109/ECTC.2001.927744

Wang, J., Ren, Z., & Nguyen, C. C. (2004). 1.156-GHz self-aligned vibrating micromechanical disk resonator. *IEEE Transactions on Ultrasonics, Ferroelectrics, and Frequency Control*, *51*(12), 1607–1628. doi:10.1109/TUFFC.2004.1386679 PMID:15690722

Ananthasuresh, G. K., Gupta, R. K., & Senturia, S. D. (1996). An approach to macromodeling of MEMS for nonlinear dynamic simulation. *Microelectromechanical Systems*, *59*, 401–407.

Belytschko, T., Krongauz, Y., Organ, D., Fleming, M., & Krysl, P. (1996). Meshless methods: An overview and recent developments. *Computer Methods in Applied Mechanics and Engineering*, *139*(1), 3–47. doi:10.1016/S0045-7825(96)01078-X

Aluru, N. R. (1999). A reproducing kernel particle method for meshless analysis of microelectromechanical systems. *Computational Mechanics*, *23*(4), 324–338. doi:10.1007004660050413

Aluru, N. R. (2000). A point collocation method based on reproducing kernel approximations. *International Journal for Numerical Methods in Engineering*, *47*(6), 1083–1121. doi:10.1002/(SICI)1097-0207(20000228)47:6<1083::AID-NME816>3.0.CO;2-N

Aluru, N. R., & Li, G. (2001). Finite cloud method: A true meshless technique based on a fixed reproducing kernel approximation. *International Journal for Numerical Methods in Engineering*, *50*(10), 2373–2410. doi:10.1002/nme.124

Li, G., & Aluru, N. R. (2003a). A boundary cloud method with a cloud-by-cloud polynomial basis. *Engineering Analysis with Boundary Elements*, *27*(1), 57–71. doi:10.1016/S0955-7997(02)00110-8

Li, G., & Aluru, N. R. (2003b). Efficient mixed-domain analysis of electrostatic MEMS. *IEEE Transactions on Computer-Aided Design of Integrated Circuits and Systems*, *22*(9), 1228–1242. doi:10.1109/TCAD.2003.816210

Nivetha, J., & Kanthamani, S. (2019). Radial Point Interpolation Method (RPIM) for the Meshless Analysis of Graphene Nano Ribbon. *Journal of the Institution of Electronics and Telecommunication Engineers*, 1–6. doi:10.1080/03772063.2018.1553639

De Los Santos, H. (2002). *RF MEMS circuit design for wireless communications*. Artech.

Liu, C., Guo, Y.-X., & Xiao, S. (2014). Capacitively loaded circularly polarized implantable patch antenna for ISM band biomedical applications. *IEEE Transactions on Antennas and Propagation, 62*(5), 2407–2417. doi:10.1109/TAP.2014.2307341

NITC. (n.d.). http://www.nitc.ac.in/electrical/ipg/pegcres/presentations/8.%20Dr.%20Ashok%20S/Dr%20ashok%20s%20.pdf

Noel, N. K., Stranks, S. D., Abate, A., Wehrenfennig, C., Guarnera, S., Haghighirad, A. A., & Petrozza, A. (2014). Lead-free organic–inorganic tin halide perovskites for photovoltaic applications. *Energy & Environmental Science, 7*(9), 3061–3068. doi:10.1039/C4EE01076K

Rashmi, S., Singh, V. K., & Puneet, K. (2018). A Compact CPW-Fed Defected Ground Microstrip Antenna for Ku Band Application. *Lecture Notes in Electrical Engineering, 443*, 231–237. doi:10.1007/978-981-10-4765-7_24

Rawat, S., & Sharma, K. K. (2014). A compact broadband microstrip patch antenna with defected ground structure for C-band applications. *Central European Journal of Engineering, 4*(3), 287–292.

Sharma, R., & Singhal, P. K. (2013). *Miniaturized Patch Antenna Loaded with Metamaterial for Microwave Sensor Quantum Matter*. Academic Press.

Singh Nikhil, Niraj, Zakir, Singh, & Bhoi. (2018). Inset Fed Circular Microstrip Antenna with Defected Ground. Lecture Notes in Electrical Engineering, 443, 605-611.

Singh, A. K., Kabeer, R. A., Shukla, M., Singh, V. K., & Ali, Z. (2013). Performance Analysis of First Iteration Koch Curve Fractal Log Periodic Antenna of Varying Angles. *Open Engineering, 3*(1), 51-57.

Singh, Saxena, Khare, Shayka, Chae, Sharma, & Bhoi. (2018). Power Harvesting through flexible rectenna at dual resonant frequency for low power devices. *International Journal of Engineering and Technology, 7*(2.33), 647-649.

Son, Y., & Heo, S. W. (2018). A novel multitarget detection algorithm for automotive FMCW radar. *2018 International Conference on Electronics, Information, and Communication (ICEIC)*, 1-3.

Mary Sindhuja, N. M., & Kanthamani, S. (2019). Three-Bit DMTL Phase Shifter for Phased Array Antennas. *Journal of Circuits, Systems, and Computers, 28*(07), 1950112. doi:10.1142/S0218126619501123

Abhishek, S., Sanjeev, G., Singh, V. K., & Bhoi, A. K. (2018). Design of Permanent Magnet Synchronous Generator for Wind Energy Conversion System. Lecture Notes in Electrical Engineering, 435, 23-32.

Energy. (n.d.). https://www.energy.gov/sites/prod/files/2017/09/f36/%5B5%5D%20NCSU%20%20Wensong%20Yu.pdf

Jamshed, A. (2015). A Novel Wide Band Microstrip-Line-Fed Antenna with Defected Ground for CP Operation. *Progress In Electromagnetics Research C*, *58*, 169–181. doi:10.2528/PIERC15052305

Kalpana, V. K. S. (2017). A Crescent Moon like Textile Antenna for C-Band Application. *International Journal of Control Theory and Application, 10*(9), 899-903.

Rachana, K., Singh, V. K., Singh, N. K., & Anurag, S. (2018). A Compact Pentagonal Textile Microstrip Antenna for Wide Band Application. Lecture Notes in Electrical Engineering, 443, 793-799.

Rebeiz, G. M. (2004). *RF MEMS: theory, design, and technology*. John Wiley & Sons.

Sharma, R. (2018). Prototype of Slotted Microstrip Patch Antenna for Multiband Application. *International Journal of Engineering & Technology, 7*(3), 1199-1201.

Singh, V. K., & Saxena, A. (2018). Two Parabolic Shape Microstrip Patch Antenna for Single Band Application. *Journal of Microwave & Technologies*, *5*(1), 33–36.

Zhu & Guo. (2016). A compact dual-band antenna for wireless body-area network applications. *IEEE Antennas Wireless Propag. Lett.*, *15*, 98–101.

Jiang, Z. H., & Werner, D. H. (2016). A compact and robust circularly polarized wearable antenna using an anisotropic metasurface. *IEEE/ACES International Conference on Wireless Information Technology and Systems and Applied Computational Electromagnetics (ACES)*. 10.1109/ROPACES.2016.7465320

Kunal, L., Archana, L., & Singh, V. K. (2018). Wide Band Triangular Patch Textile Antenna with Partial Ground Plane. *Advances in Intelligent Systems and Computing*, *632*, 577–583. doi:10.1007/978-981-10-5520-1_52

Lohia, Singh, & Saxena. (2020). A Flexible Multi slot wireless Antenna designed for Reconfigurable Frequency. *2nd International Conferences on Emerging Trends and Advances in Electrical Engineering and Renewable Energy (ETAEERE-2020)*.

Ronan, E. R., Sudhoff, S. D., Glover, S. F., & Galloway, D. L. (2000, February). Application of power electronics to the distribution transformer. In *APEC 2000. Fifteenth Annual IEEE Applied Power Electronics Conference and Exposition (Cat. No. 00CH37058)* (Vol. 2, pp. 861-867). IEEE. 10.1109/APEC.2000.822606

Saxena, Singh, Kumar, & Ansari. (2018). A Concentric Double hole Shape Microstrip Patch Antenna at 3.12 GHz Resonant Frequency. In *Advanced Computation and Telecommunication (ICACAT)*. IEEE.

Sharma, Yadav, & Singh. (2018). Design of Circularly Polarized Antenna with Different Iterations for UWB Applications. Lecture Notes in Electrical Engineering, 443, 441-447.

Xiaomu, H., Yan, S., & Vandenbosch, G. A. E. (2007). Wearable button antenna for dual-band WLAN applications with combined on and offbody radiation patterns. *IEEE Transactions on Antennas and Propagation*, *65*(3), 1384–1387. doi:10.1109/TAP.2017.2653768

Fang, Q., Lee, S. Y., Parmana, H., Ghorbani, K., & Cosic, I. (2011). Developing a wireless implantable body sensor network in MICS band. *IEEE Transactions on Information Technology in Biomedicine*, *15*(4), 567–576. doi:10.1109/TITB.2011.2153865 PMID:21571615

Khare, Bhoi, Sharma, & Lohia. (2019). *Missile Structured Wearable Antenna for Power Harvesting Application: Emerging Research and Opportunities*. Academic Press.

Lucyszyn, S. (2004). Review of radio frequency microelectromechanical systems technology. *IEE Proceedings. Science Measurement and Technology*, *151*(2), 93–103. doi:10.1049/ip-smt:20040405

Princi, T., Upendra, T., Singh, V. K., & Bhoi, A. K. (2018). Solution of Economic Load Dispatch Problems through Moth Flame Optimization Algorithm. *Lecture Notes in Electrical Engineering*, *462*, 287–294. doi:10.1007/978-981-10-7901-6_31

Saxena, A., & Singh, V. K. (2019a). Antenna for S-Band Application Utilizing Flexible Material. *International Journal of Broadband Cellular Communication.*, *5*(1), 1–4p.

Shakib & Moghavvemi. (2017). Design of a tri-band off-body antenna for WBAN communication. *IEEE Antennas Wireless Propag. Lett.*, *16*, 210–213.

Swathi, G. (2018). Design an Efficient Power Electronics Transformer (PET) for the Improvement of Quality in Power. *IACSIT International Journal of Engineering and Technology*, *7*(3.12), 1046. doi:10.14419/ijet.v7i3.12.17629

Xu & Li. (2012). A Dual band Microstrip Antenna for Wearable Application. *IEEE (ISAPE) Conference Proceeding*, 109-112.

Antoskiewicz, K., & Shafai, L. (2017). Impedance characteristics of circular microstrip patches. *IEEE Transactions on Antennas and Propagation*, *38*(6), 942–946. doi:10.1109/8.55597

Duan, Z., Guo, Y. X., Xue, R. F., Je, M., & Kwong, D. L. (2012). Differentially fed dual band implantable antenna for biomedical applications. *IEEE Transactions on Antennas and Propagation*, *60*(12), 5587–5595. doi:10.1109/TAP.2012.2209197

Niharika, S., Singh, V. K., Rahul, S., & Prakash, S. J. (2018). Microstrip Textile Antenna with Jeans Substrate with Applications in S-Band. Lecture Notes in Electrical Engineering, 462, 369-376.

Rebeiz, G. M., & Muldavin, J. B. (2001). RF MEMS switches and switch circuits. *IEEE Microwave Magazine*, *2*(4), 59–71. doi:10.1109/6668.969936

Saxena, A., & Singh, V. K. (2019b). Triangle Cut Circle Shape Bending Antenna for Wide Band Applications. *Journal of Microwave Engineering and Technology.*, *6*(2), 14–17p.

Singh Nikhil, Sharma Niraj, Zakir, Singh, & Bhoi. (2018). Inset Fed Circular Microstrip Antenna with Defected Ground. Lecture Notes in Electrical Engineering, 443, 605-611.

Wang, D., Mao, C., & Li, J. (2007). Auto Balancing Electronic Power Transformer. *Zhongguo Dianji Gongcheng Xuebao*, *27*(6), 77–83.

Yadav, Singh, & Mohan. (2019). *Design of a U-shaped Circularly wearable antenna with DGS on a fabric substrate for WLAN and C-band Applications*. Academic Press.

Agilent, H. S. M. S. (2005). *-285x Series Surface Mount Zero Bias Schottky Detector Diodes-Data Sheet 5898-4022EN*. Agilent Technologies, Inc.

Kiourti, A., & Nikita, K. S. (2012). Miniature scalp implantable antennas for telemetry in MICS and ISM bands: Design, safety considerations and link budget analysis. *IEEE Transactions on Antennas and Propagation*, *60*(8), 3568–3575. doi:10.1109/TAP.2012.2201078

Manju, D., & Singh, V. K. (2018). Antenna for Wireless Area Network and Bluetooth Application. Lecture Notes in Electrical Engineering, 462, 377-384.

Prabhakaran, S., Stephens, C., Johnson, F., & Lorio, L. (2008). High frequency power transformers for future military propulsion and energy systems. *Proc. Electron. Mach. Technol. Symp.*, 1–21.

Pranonsatit, S., Holmes, A. S., Robertson, I. D., & Lucyszyn, S. (2006). Single-pole eight-throw RF MEMS rotary switch. *Journal of Microelectromechanical Systems*, *15*(6), 1735–1744. doi:10.1109/JMEMS.2006.883578

Rachana, K., Singh, V. K., Singh, N. K., & Anurag, S. (2018). A Compact Pentagonal Textile Microstrip Antenna for Wide Band Application. Lecture Notes in Electrical Engineering, 443, 793-799.

Simorangir, Yang, Esselle, & Diao. (2017). A varactor tuned frequency reconfigurable fabric antenna embedded in polymer. *IEEE MTT-S International microwave Symposium (IMS)*, 204-207.

Singh, Saxena, Khare, Shayka, Chae, Sharma, & Bhoi. (2018). Power Harvesting through flexible rectenna at dual resonant frequency for low power devices. *International Journal of Engineering and Technology, 7*(2.33), 647-649.

Hyman, D., Lam, J., Warneke, B., Schmitz, A., Hsu, T. Y., Brown, J., ... Lee, J. (1999). Surface-micromachined RF MEMs switches on GaAs substrates. *International Journal of RF and Microwave Computer-Aided Engineering, 9*(4), 348-361.

Jain, S., Singh, V. K., & Ayub, S. (2013). Band Width and Gain Optimization of a Wide Band Gap Coupled Patch Antenna. *International Journal of Engineering Sciences & Research Technology*, 649-652.

Kunov, G. (2014, May). Matlab-Simulink model of solid-state transformer realized with matrix converters. In *2014 18th International Symposium on Electrical Apparatus and Technologies (SIELA)* (pp. 1-4). IEEE. 10.1109/SIELA.2014.6871865

Naresh, B., & Singh, V. K. (2020). Integration of RF Rectenna with thin film solar cell to power wearable electronics. *International Journal of Microwave and Wireless Technologies*.

Rahul, S., & Singh, V. K. (2018). Multi Resonant Textile Antenna with Partial Ground for Multiband Applications. Lecture Notes in Electrical Engineering, 462, 359-367.

Roundy, S. J. (2003). *Energy scavenging for wireless sensor nodes with a focus on vibration to electricity conversion* (PhD Thesis). University of California, Berkeley, CA.

Dujic, D., Zhao, C., Mester, A., Steinke, J. K., Weiss, M., Schmid, S. L., Chaudhuri, T., & Stefanutti, P. (2013, December). Power electronic traction transformer: Low voltage prototype. *IEEE Transactions on Power Electronics*, *28*(12), 5522–5534. doi:10.1109/TPEL.2013.2248756

Pathak, R. S., Singh, V. K., & Ayub, S. (2012). Dual band Microstrip Antenna for GPS/ WLAN/ WiMax Applications. *International Journal of Emerging Trends in Engineering and Development, 7*(2), 154-159.

Saxena. (2019). A La Shape Antenna for High Frequencies Application. In *Design and Optimization of Sensors and Antennas for Wearable Devices: Emerging Research and Opportunities*. Academic Press.

Shukla, S. S., Verma, R. K., & Gohir, G. S. (2015). Investigation of the effect of Substrate material on the performance of Microstrip antenna. *4th International Conference on Reliability, Infocom Technologies and Optimization*, 1-3. 10.1109/ICRITO.2015.7359350

Xu & Li. (2012). A Dual band Microstrip Antenna for Wearable Application. *IEEE (ISAPE) Conference Proceeding*, 109-112.

Yao, J. J., & Chang, M. C. (1995, December). Surface micromachined miniature switch for telecommunications applications with signal frequencies from DC up to 4 GHZ. *Proceedings of the 1995 8th International Conference on Solid-State Sensors and Actuators and Eurosensors IX. Part 1 (of 2)*. 10.1109/SENSOR.1995.721827

Abhishek, S., Sanjeev, G., Singh, V. K., & Bhoi, A. K. (2018). Design of Permanent Magnet Synchronous Generator for Wind Energy Conversion System. Lecture Notes in Electrical Engineering, 435, 23-32.

Hall, P. S., & Hao, Y. (2006). *Antennas and Propagation for Body-centric Wireless Communications*. Artech House.

Hyman, D., & Mehregany, M. (1998, October). Contact physics of gold microcontacts for MEMS switches. In *Electrical Contacts-1998. Proceedings of the Forty-Fourth IEEE Holm Conference on Electrical Contacts (Cat. No. 98CB36238)* (pp. 133-140). IEEE. 10.1109/HOLM.1998.722438

Manjrekar, M. D., Kieferndorf, R., & Venkataramanan, G. (2000, October). Power electronic transformers for utility applications. In *Conference Record of the 2000 IEEE Industry Applications Conference. Thirty-Fifth IAS Annual Meeting and World Conference on Industrial Applications of Electrical Energy (Cat. No. 00CH37129)* (Vol. 4, pp. 2496-2502). IEEE. 10.1109/IAS.2000.883173

Poornima, S. (2018). *Design and Analysis of Microstrip Antenna Using Multilayer Feed-Forward Back-Propagation Neural Network (MLPFFBP-ANN). Lecture Notes in Electrical Engineering*, 462.

Saxena & Khare. (2019). A Hammer type Textile Antenna with partial circle ground for wide band applications. In *Design and Optimization of Sensors and Antennas for Wearable Devices: Emerging Research and Opportunities*. Academic Press.

Singh, Ali, Singh, & Ayub. (2013). Dual Band Microstrip Antenna for UMTS/WLAN/WIMAX Applications. *International Conference on Communication Systems and Network Technology (CSNT-2013),* 47 – 50.

Ashish, K., Archana, L., Singh, V. K., & Bhoi, A. K. (2018). Estimation of Frequency Band of Microstrip Antenna (MSA) with Radial Basis Function (RBF). *Lecture Notes in Electrical Engineering*, *462*, 385–391. doi:10.1007/978-981-10-7901-6_42

Chen, Z., Guerrero, J. M., & Blaabjerg, F. (2006, June). A review of state of art power electronics for wind turbines. *IEEE Transactions on Industrial Electronics*, *53*(4), 1002–1016.

Goldsmith, C. L., Yao, Z., Eshelman, S., & Denniston, D. (1998). Performance of low-loss RF MEMS capacitive switches. *IEEE Microwave and Guided Wave Letters*, *8*(8), 269–271. doi:10.1109/75.704410

Singh, V. K., Ali, Z., Ayub, S., & Singh, A. K. (2014b). A wide band Compact Microstrip Antenna for GPS/DCS/PCS/WLAN Applications. In Intelligent Computing, Networking, and Informatics. Academic Press.

Agrawal, N. (2020). *Performance Analysis of Cascade H-Bridge Multilevel Inverter Topology With Filter Circuit and Without Filter Circuit. In Design and Optimization of Sensors and Antennas for Wearable Devices: Emerging Research and Opportunities*. IGI Global. doi:10.4018/978-1-5225-9683-7.ch008

Forest Survey of India. (2011). Government of India.

Goesmann, H., & Feldmann, C. (2010). Nanoparticulate functional materials. *Angewandte Chemie International Edition*, *49*(8), 1362–1395. doi:10.1002/anie.200903053 PMID:20108287

Haykin, S. (2003). *Neural Networks* (2nd ed.). PHI.

Ouyang, Y., & Chappell, W. J. (2008, February). High Frequency Properties of Electro-Textiles for Wearable Antenna Applications. *IEEE Transactions on Antennas and Propagation*, *56*(2), 381–389. doi:10.1109/TAP.2007.915435

Ronan, E. R., Sudhoff, S. D., Glover, S. F., & Galloway, D. L. (2002, April). A power electronic-based distribution transformer. *IEEE Transactions on Power Delivery*, *17*(2), 537–543. doi:10.1109/61.997934

Sarkar, T. K., & Palma, M. S. (2014). A History of the evolution of RADAR. *44th European Microwave Conference*, 734-737. 10.1109/EuMC.2014.6986539

Saxena. (2019). A La Shape Antenna for High Frequencies Application. In *Design and Optimization of Sensors and Antennas for Wearable Devices: Emerging Research and Opportunities.* IGI Global.

Sharma, Ashish, & Vinod. (2018). Design of Circularly Polarized Antenna with Different Iterations for UWB Applications. Lecture Notes in Electrical Engineering, 443, 441-447.

Simorangkir, Yang, & Esselle. (2018). Method to Realize Robust Flexible Electronically Tunable Antennas Using Polymer Embedded Conductive Fabric. IEEE Transactions on Antennas and Propagation, 66.

Tsoulos, G. V. (2000). *Adaptive antennas for wireless communications*. Wiley-IEEE Press.

Yadav, A., Singh, V.K., Yadav, P., Beliya, A.K., Bhoi, A.K., & Barsocchi, P. (2020). Design of Circularly Polarized Triple-Band Wearable Textile Antenna with Safe Low SAR for Human Health. *Electronics*, *9*(1366).

Backlund, B., Rahimo, M., Klaka, K., & Siefken, J. (2009). Topologies, voltage ratings, and state of the art high power semiconductor devices for medium voltage wind energy system. *Proc. IEEE PEMWA*, 1–6.

Rahul, Singh, Singh, Prakash, & Bhoi. (2018). Multi Resonant Textile Antenna with Partial Ground for Multiband Applications. Lecture Notes in Electrical Engineering, 462, 359-367.

Singh, Ali, & Ayub. (2010). Design of compact triple band microstrip antenna for wireless communication. International Journal of Electronics & Communication Engineering, 3(1), 323-330.

Sovero, E. A., Mihailovich, R., Deakin, D. S., Higgins, J. A., Yao, J. J., DeNatale, J. F., & Hong, J. H. (1999, August). Monolithic GaAs PHEMT MMICs integrated with high performance MEMS microrelays. In *1999 SBMO/IEEE MTT-S International Microwave and Optoelectronics Conference* (Vol. 1, pp. 257-260). IEEE. 10.1109/IMOC.1999.867103

Srivastava, R., Singh, V. K., & Ayub, S. (2015). Comparative Analysis and Bandwidth Enhancement with Direct Coupled C Slotted Microstrip Antenna for Dual Wide Band Applications. In Frontiers of Intelligent Computing: Theory and Applications. Springer.

Deepak & Pathak. (2013). A study on inverted T shaped micro strip antenna at different frequencies. *International Journal of Engineering and Computer Science, 2*(11), 3180-3183.

Hosaka, H., Kuwano, H., & Yanagisawa, K. (1994). Electromagnetic microrelays: Concepts and fundamental characteristics. *Sensors and Actuators. A, Physical*, *40*(1), 41–47. doi:10.1016/0924-4247(94)85028-3

Kolar, J. W., & Ortiz, G. (2014, May). Solid-state-transformers: Key components of future traction and smart grid systems. *International Power Electronics Conference (IPEC), Hi-roshima.*

Lumba, S., Singh, V. K., & Srivastava, R. (2014). Bandwidth Enhancement by Direct Coupled Antenna for WLAN/GPS/WiMax Applications & Feed Point Analysis through ANN. In Computational Intelligence in Data Mining. Academic Press.

Niharika, S., Singh, V. K., Rahul, S., & Prakash, S. J. (2018). Microstrip Textile Antenna with Jeans Substrate with Applications in S-Band. Lecture Notes in Electrical Engineering, 462, 369-376.

Abhishek, S., Sanjeev, G., Singh, V. K., & Bhoi, A. K. (2018). Design of Permanent Magnet Synchronous Generator for Wind Energy Conversion System. Lecture Notes in Electrical Engineering, 435, 23-32.

Jiang, Z. H., Brocker, D. E., Sieber, P. E., & Werner, D. H. (2014). A compact, low-profile metasurface-enabled antenna for wearable medical body-area network devices. *IEEE Transactions on Antennas and Propagation*, *62*(8), 402–4030. doi:10.1109/TAP.2014.2327650

Komrska, T., & Peroutka, Z. (2008). Main traction converter with medium frequency transformer: Control of converters around MF transformer. *Proc. SPEEDAM*, 1194–1198. 10.1109/SPEEDHAM.2008.4581256

Manju, D., & Singh, V. K. (2018). Antenna for Wireless Area Network and Bluetooth Application. Lecture Notes in Electrical Engineering, 462, 377-384.

Sharma, P., Yadav, A., & Singh, V. K. (2018). Design of Circularly Polarized Antenna with Different Iterations for UWB Applications. In A. Kalam, S. Das, & K. Sharma (Eds.), *Advances in Electronics, Communication and Computing. Lecture Notes in Electrical Engineering* (Vol. 443, pp. 441–447). Springer. doi:10.1007/978-981-10-4765-7_47

Taylor, W. P., Allen, M. G., & Dauwalter, C. R. (1997). Batch fabricated electromagnetic microrelays. In Proceedings of the relay conference-national association of relay manufacturers (vol. 45, pp. 8-1). National Association of Relay Manufacturers.

Jain, S., Singh, V. K., & Ayub, S. (2013). Band Width and Gain Optimization of a Wide Band Gap Coupled Patch Antenna. *International Journal of Engineering Sciences & Research Technology*, 649-652.

Singh, N. K., Sharma, N., Ali, Z., Singh, V. K., & Bhoi, A. K. (2018). Inset Fed Circular Microstrip Antenna with Defected Ground. In A. Kalam, S. Das, & K. Sharma (Eds.), *Advances in Electronics, Communication and Computing. Lecture Notes in Electrical Engineering* (Vol. 443, pp. 605–611). Springer. doi:10.1007/978-981-10-4765-7_63

Tilmans, H. A., Fullin, E., Ziad, H., Van de Peer, M. D., Kesters, J., Van Geffen, E., . . . Naso, F. (1999, January). A fully-packaged electromagnetic microrelay. In *Technical Digest. IEEE International MEMS 99 Conference. Twelfth IEEE International Conference on Micro Electro Mechanical Systems (Cat. No. 99CH36291)* (pp. 25-30). IEEE. 10.1109/MEMSYS.1999.746746

Yadav, A., Kumar Singh, V., Kumar Bhoi, A., Marques, G., Garcia-Zapirain, B., & de la Torre Díez, I. (2020). Wireless Body Area Networks: UWB Wearable Textile Antenna for Telemedicine and Mobile Health Systems. *Micromachines*, *11*(6), E558. doi:10.3390/mi11060558

Kushwaha, R., Singh, V. K., Singh, N. K., Saxena, A., & Sharma, D. (2018). A Compact Pentagonal Textile Microstrip Antenna for Wide Band Application. In A. Kalam, S. Das, & K. Sharma (Eds.), *Advances in Electronics, Communication and Computing. Lecture Notes in Electrical Engineering* (Vol. 443, pp. 793–799). Springer. doi:10.1007/978-981-10-4765-7_82

Pathak, R. S., Singh, V. K., & Ayub, S. (2012). Dual band Microstrip Antenna for GPS/ WLAN/ WiMax Applications. *International Journal of Emerging Trends in Engineering and Development, 7*(2), 154-159.

She, X., & Huang, A. (2013, July). *Solid state transformer in the future smart electrical system. In 2013 IEEE Power & Energy Society General Meeting*. IEEE.

Wright, J. A., & Tai, Y. C. (1999, January). Magnetostatic MEMS relays for the miniaturization of brushless DC motor controllers. In *Technical Digest. IEEE International MEMS 99 Conference. Twelfth IEEE International Conference on Micro Electro Mechanical Systems (Cat. No. 99CH36291)* (pp. 594-599). IEEE. 10.1109/MEMSYS.1999.746895

Yadav & Singh. (2019). Design of U-shape with DGS circularly polarized wearable antenna on fabric substrate for WLAN and C-Band applications. *Journal of Computational Electronics, 18*(3), 1103–1109. . doi:10.100710825-019-01342-2

Gretillat, M. A., Gretillat, F., & De Rooij, N. F. (1999). Micromechanical relay with electrostatic actuation and metallic contacts. *Journal of Micromechanics and Microengineering, 9*(4), 324.

Niharika, S., Singh, V. K., Rahul, S., & Prakash, S. J. (2018). Microstrip Textile Antenna with Jeans Substrate with Applications in S-Band. Lecture Notes in Electrical Engineering, 462, 369-376.

Singh, Ali, Singh, & Ayub. (2013). Dual Band Microstrip Antenna for UMTS/WLAN/WIMAX Applications. *International Conference on Communication Systems and Network Technology (CSNT-2013),* 47 – 50.

Singh, V. K., Dhupkariya, S., & Bangari, N. (2017). Wearable Ultra Wide Dual Band Flexible Textile Antenna for WiMax/WLAN Application. *International Journal of Wireless Personal Communications, 95*(2), 1075–1086. doi:10.100711277-016-3814-7

Manju, D., & Singh, V. K. (2018). Antenna for Wireless Area Network and Bluetooth Application. Lecture Notes in Electrical Engineering, 462, 377-384.

Sanders, C. (1998). *MCNC thermally actuated microrelays*. News Release.

Singh, N. K., Singh, V. K., Saxena, A., Bhoi, A. K., Garg, A., & Sherpa, K. S. (2018). A Compact Slotted Textile Patch Antenna for Ultra-wide Band Application. Lecture Notes in Electrical Engineering, 443, 53-59. doi:10.1007/978-981-10-4765-7_6

Singh, N., Singh, A. K., & Singh, V. K. (2015). Design & Performance of Wearable Ultra Wide Band Textile Antenna for Medical Applications. *Microwave and Optical Technology Letters, 57*(7), 1553-1557. doi:10.1002/mop.29131

Jain, S., Singh, V. K., & Ayub, S. (2013). Band Width and Gain Optimization of a Wide Band Gap Coupled Patch Antenna. *International Journal of Engineering Sciences & Research Technology*, 649-652.

Larson, L. E., Hackett, R. H., & Lohr, R. F. (1991, June). Microactuators for GaAs-based microwave integrated circuits. In *TRANSDUCERS'91: 1991 International Conference on Solid-State Sensors and Actuators. Digest of Technical Papers* (pp. 743-746). IEEE. 10.1109/SENSOR.1991.148990

Singh, Ali, & Ayub. (2010). Design of compact triple band microstrip antenna for wireless communication. *International Journal of Electronics & Communication Engineering, 3*(1), 323-330.

Yu, X. J., Tang, Q., Liang, M., & Xin, H. (2014). Study of power efficiency of non-foster impedance matching for electrically small antenna. *IEEE Conf Publ.* 10.1109/USNC-URSI.2014.6955512

Deepak & Pathak. (2013). A study on inverted T shaped micro strip antenna at different frequencies. *International Journal of Engineering and Computer Science, 2*(11), 3180-3183.

Hu, B., Gao, G. P., He, L. L., Cong, X. D., & Zhao, J. N. (2016). Bending and on-arm effects on a wearable antenna for 2.45 GHz body area network. *Antennas Wireless Propag IEEE Letter, 15*, 378–381. doi:10.1109/LAWP.2015.2446512

Pathak, R. S., Singh, V. K., & Ayub, S. (2012). Dual band Microstrip Antenna for GPS/ WLAN/ WiMax Applications. *International Journal of Emerging Trends in Engineering and Development, 7*(2), 154-159.

Saini, R., Singh, V. K., Singh, N., Saini, J. P., & Bhoi, A. K. (2018). Multi Resonant Textile Antenna with Partial Ground for Multiband Applications. Lecture Notes in Electrical Engineering, 462, 359-367. doi:10.1007/978-981-10-7901-6_39

De Los Santos, H. J., Kao, Y. H., Caigoy, A. L., & Ditmars, E. D. (1997, February). *Microwave and mechanical considerations in the design of MEM switches for aerospace applications. In 1997 IEEE Aerospace Conference* (Vol. 3). IEEE.

Dubey, R., Singh, V. K., Bhoi, A. K., & Ali, Z. (2018). Realization of trophy shaped flexible wearable antenna based on foam substrate. *International Journal of Engineering & Technology, 7*, 222-224.

Loni, Singh, Tripathi, Chae, Sharma, & Bhoi. (2018). Improvement of the performance of wearable textile antenna. *IACSIT International Journal of Engineering and Technology.*

Singh, Ali, Singh, & Ayub. (2013). Dual Band Microstrip Antenna for UMTS/WLAN/WIMAX Applications. *International Conference on Communication Systems and Network Technology (CSNT-2013)*, 47 – 50.

Agrawal, Singh, & Bansal. (2017). A Multilevel Inverter Topology using Reverse-Connected Voltage Sources. *International Conference on Energy, Communication, Data Analytics and soft Computing (ICECDS 2017)*, 1290-1295.

Axelsson, S. R. J. (2007, April). Analysis of Random Step Frequency Radar and Comparison With Experiments. *IEEE Transactions on Geoscience and Remote Sensing, 45*(4), 890–904. doi:10.1109/TGRS.2006.888865

Bangari, N., Singh, V. K., & Vanga, B. (2018). Dual-Band Wearable Rectenna for Low-Power RF Energy Harvesting. Lecture Notes in Electrical Engineering, 436, 13-21.

District Forest Office Dantewada. (n.d.). Government of India.

Hassoun, M. H. (1999). *Fundamentals of Artificial Neural Network*. Prentice Hall of India.

Lemey, S., Declercq, F., & Rogier, H. (2014, November). Textile Antennas as Hybrid Energy-Harvesting Platforms. *Proceedings of the IEEE, 102*(11), 1833–1857. doi:10.1109/JPROC.2014.2355872

Matthews, J. G. G., & Pettitt, G. (2009). Development of flexible, wearable antennas. *3rd European Conference on Antenna and Propagation.*

Mourad, M., & Essaaidi, M. (2014). A dual ultra wide band slotted antenna for C and X-band application. *Progress in Electromagnetics Research Letters*, *47*, 91–96. doi:10.2528/PIERL14052103

Peroulis, D. (2003). *RF MEMS devices for multifunctional integrated circuits and antennas* (Doctoral dissertation). University of Michigan.

Qin, H. S., & Kimball, J. W. (2009). AC-AC dual active bridge converter for solid state transformer. *Proc. IEEE ECCE*, 3039–3044.

Tong, X., Liu, C., Liu, X., Guo, H., & Yang, X. (2018), Switchable ON-/OFF-Body Antenna for 2.45 GHz WBAN Applications. IEEE Transactions on Antennas and Propagation, 66(2).

Verma, A., Diwakar, A. K., & Patel, R.P. (2019). Synthesis and Characterization of High-Performance Solar Cell. *International Journal of Scientific Research in Physics and Applied Sciences, 7*(2).

Yadav, A., Kumar Singh, V., Kumar Bhoi, A., Marques, G., Garcia-Zapirain, B., & de la Torre Díez, I. (2020). Wireless Body Area Networks: UWB Wearable Textile Antenna for Telemedicine and Mobile Health Systems. *Micromachines, 11*(6), E558.

Bano, N., Singh, V. K., Bhoi, A. K., & Sharma, S. (2018). Design and Analysis of Key shaped wearable textile antenna. *International Journal of Engineering & Technology, 7*, 219-221.

Saxena, A., & Singh, V. K. (2018). A Moon-Strip Line Antenna for Multi-Band Applications at 5.44 GHz Resonant Frequency. *Fourth International Conference on Advances in Electrical, Electronics, Information, Communication and Bio-Informatics (AEEICB).* 10.1109/AEEICB.2018.8480998

Majumder, S., McGruer, N. E., Zavracky, P. M., Adams, G. G., Morrison, R. H., & Krim, J. (1997, June). Measurement and modelling of surface micromachined, electrostatically actuated microswitches. In *Proceedings of International Solid State Sensors and Actuators Conference (Transducers' 97)* (Vol. 2, pp. 1145-1148). IEEE. 10.1109/SENSOR.1997.635407

Saxena, Singh, Lohia, & Faisal. (2020). A Jawbreaker-Shaped Fragmentary Ground Textile Flexible Antenna. *International Journal of Electronics, Communications, and Measurement Engineering.*

Singh, Ali, & Ayub. (2010). Design of compact triple band microstrip antenna for wireless communication. *International Journal of Electronics & Communication Engineering, 3*(1), 323-330.

Singh, Singh, Kumar, Yadav, & Bhoi. (2018). Rectenna design for electromagnetic energy harvesting and wireless power transfer. *International Journal of Engineering & Technology, 7*, 632-633.

Deepak & Pathak. (2013). A study on inverted T shaped micro strip antenna at different frequencies. *International Journal of Engineering and Computer Science, 2*(11), 3180-3183.

Schlaak, H. F. (1997). Silicon-microrelay-a small signal relay with electrostatic actuator. *Proc. 45th relay conf.*

Singh, Saxena, Khare, Shakya, Sharma, & Bhoi. (2018). Power harvesting through flexible rectenna at dual resonant frequency for low power devices. *International Journal of Engineering & Technology, 7*, 647-649.

Singh, V. K., & Singh, N. K. (2018). Rectenna design for electromagnetic energy harvesting and wireless power transfer. *IACSIT International Journal of Engineering and Technology.*

Suzuki, K., Chen, S., Marumoto, T., Ara, Y., & Iwata, R. (1999, June). A micromachined RF microswitch applicable to phased-array antennas. In *1999 IEEE MTT-S International Microwave Symposium Digest (Cat. No. 99CH36282)* (Vol. 4, pp. 1923-1926). IEEE. 10.1109/MWSYM.1999.780350

Muldavin, J. B., & Rebeiz, G. M. (1999, June). 30 GHz tuned MEMS switches. In *1999 IEEE MTT-S International Microwave Symposium Digest (Cat. No. 99CH36282)* (Vol. 4, pp. 1511-1514). IEEE. 10.1109/MWSYM.1999.780241

Saxena & Singh. (2018). A Moon-Strip Line Antenna for Multi-Band Applications at 5.44 GHz Resonant Frequency. *2018 Fourth International Conference on Advances in Electrical, Electronics, Information, Communication and Bio-Informatics (AEEICB).*

Singh, Singh, & Singh. (2015). Design & Performance of Wearable Ultra Wide Band Textile Antenna for Medical Applications. *Microwave and Optical Technology Letters, 57*(7), 1553-1557.

Yao, Z. J., Chen, S., Eshelman, S., Denniston, D., & Goldsmith, C. (1999). Micromachined low-loss microwave switches. *Journal of Microelectromechanical Systems, 8*(2), 129–134. doi:10.1109/84.767108

Goldsmith, C., Randall, J., Eshelman, S., Lin, T. H., Denniston, D., Chen, S., & Norvell, B. (1996, June). Characteristics of micromachined switches at microwave frequencies. In *1996 IEEE MTT-S International Microwave Symposium Digest. IEEE MTT-S International Microwave Symposium, 2*, 1141–1144.

Saxena, Singh, Lohia, & Faisal. (2020). A Jawbreaker-Shaped Fragmentary Ground Textile Flexible Antenna. *International Journal of Electronics, Communications, and Measurement Engineering.*

Komura, Y., Sakata, M., Seki, T., Kobayashi, K., Sano, K., Horiike, S., & Ozawa, K. (1999, April). Micro machined relay for high frequency application. In Proceedings Of The Relay Conference-National Association Of Relay Manufacturers (Vol. 47, pp. 12-12). Academic Press.

Boopathi, K. M., Ramesh, M., Perumal, P., Huang, Y. C., Tsao, C. S., Chen, Y. F., & Chu, C. W. (2015). Preparation of metal halide perovskite solar cells through a liquid droplet assisted method. *Journal of Materials Chemistry. A, Materials for Energy and Sustainability*, *3*(17), 9257–9263. doi:10.1039/C4TA06392A

Liu, Y., Meng, H., Zhang, H., & Wang, X. (2008). Motion compensation of moving targets for high range resolution stepped-frequency radar. *Sensors (Basel)*, *8*(5), 3429–3437. doi:10.33908053429 PMID:27879886

Masood, R., Person, C., & Sauleau, R. (2007). A dual-mode, dual-port pattern diversity antenna for 2.45-GHz WBAN. *IEEE Antennas and Wireless Propagation Letters*, *16*, 1064–1067. doi:10.1109/LAWP.2016.2620724

Nguyen, C. C., Katehi, L. P., & Rebeiz, G. M. (1998). Micromachined devices for wireless communications. *Proceedings of the IEEE*, *86*(8), 1756–1768. doi:10.1109/5.704281

Saxena & Khare. (2019). A Hammer type Textile Antenna with partial circle ground for wide band applications. In *Design and Optimization of Sensors and Antennas for Wearable Devices: Emerging Research and Opportunities*. IGI Global.

She, X., Huang, A. Q., & Burgos, R. (2013). Review of solid-state transformer technologies and their application in power distribution systems. *IEEE Journal of Emerging and Selected Topics in Power Electronics*, *1*(3), 186–198. doi:10.1109/JESTPE.2013.2277917

Singh Nikhil, Niraj, Zakir, Singh, & Bhoi. (2018). Inset Fed Circular Microstrip Antenna with Defected Ground. Lecture Notes in Electrical Engineering, 443, 605-611.

Singh, V. K., Lala, A., & Singh, A. K. (2020). Novel Inset Feed Circular Slotted Microstrip Antenna Using Multilayer Feed-Forward Back-Propagation and Radial Basis Function Neural Network. *National Academy Science Letters*.

Solis, A., Whittow, W. G., Alexandridis, A. A., & Vardaxoglou, J. C. (2014). Embroidery and Related Manufacturing Techniques for Wearable Antennas: Challenges and Opportunities. *Electronics (Basel)*, *3*(2), 314–338. doi:10.3390/electronics3020314

Tapas, V., & Sharma, P. (2014). The Potentials of Solar Drying in Bastar Region of Chhattisgarh State in India. *International Congress on Renewable Energy (ICORE-2014)*.

Thiyagarajan, V., Somasundaram, P., & Ramash Kumar, K. (2019, June 15). Simulation and Analysis of Novel Extendable Multilevel Inverter Topology. *Journal of Circuits, Systems, and Computers*, *28*(6), 1950089. Advance online publication. doi:10.1142/S0218126619500890

Zeland Software Inc. (n.d.). *IE3D: MoM-based EM simulator*. http:// www.zeland.com

Admschik, M., Ertl, S., Schmid, P., Gluche, P., Flöter, A., & Kohn, E. (1999, June). Electrostatic diamond micro switch. In *Tech. Digest, 10th Int. Conf. on Solid-State Sensors and Actuators* (pp. 1284-7). Academic Press.

Gretillat, M. A., Gretillat, F., & De Rooij, N. F. (1999). Micromechanical relay with electrostatic actuation and metallic contacts. *Journal of Micromechanics and Microengineering*, *9*(4), 324–331. doi:10.1088/0960-1317/9/4/307

Saxena, Singh, Lohia, & Faisal. (2020). A Jawbreaker-Shaped Fragmentary Ground Textile Flexible Antenna. *International Journal of Electronics, Communications, and Measurement Engineering*.

Simon, J., Saffer, S., & Kim, C. J. (1996, February). A micromechanical relay with a thermally-driven mercury micro-drop. In *Proceedings of Ninth International Workshop on Micro Electromechanical Systems* (pp. 515-520). IEEE. 10.1109/MEMSYS.1996.494035

Sun, X. Q., Farmer, K. R., & Carr, W. N. (1998, January). A bistable microrelay based on two-segment multimorph cantilever actuators. In *Proceedings MEMS 98. IEEE. Eleventh Annual International Workshop on Micro Electro Mechanical Systems. An Investigation of Micro Structures, Sensors, Actuators, Machines and Systems (Cat. No. 98CH36176* (pp. 154-159). IEEE.

Kruglick, E. J., & Pister, K. S. (1998, June). *Bistable MEMS relays and contact characterization. 1998 Solid-State Sensor and Actuator Workshop.*

Schiele, I., & Hillerich, B. (1999). Comparison of lateral and vertical switches for application as microrelays. *Journal of Micromechanics and Microengineering*, *9*(2), 146–150. doi:10.1088/0960-1317/9/2/310

Kruglick, E. J., & Pister, K. S. (1999). Lateral MEMS microcontact considerations. *Journal of Microelectromechanical Systems*, *8*(3), 264–271. doi:10.1109/84.788630

Varadan, V. K., Vinoy, K. J., & Jose, K. A. (2003). *RF MEMS and their Applications. Pennsylvania State University*. Wiley Online Library.

Iannacci, J. (2017). RF-MEMS for high-performance and widely reconfigurable passive components–A review with focus on future telecommunications, Internet of Things (IoT) and 5G applications. *Journal of King Saud University. Science*, *29*(4), 436–443. doi:10.1016/j.jksus.2017.06.011

Electronicsforu. (n.d.). https://electronicsforu.com/resources/learn-electronics/solid-state-transformers-power-grid

Im, J. H., Chung, J., Kim, S. J., & Park, N. G. (2012). Synthesis, structure, and photovoltaic property of a nanocrystalline 2H perovskite-type novel sensitizer (CH 3 CH 2 NH 3) PbI 3. *Nanoscale Research Letters*, *7*(1), 353. doi:10.1186/1556-276X-7-353 PMID:22738298

Kumar, & Ray. (2003). *Broadband Microstrip antennas*. Artech House.

Lim, K., Pinel, S., Davis, M., Sutono, A., Lee, C. H., Heo, D., & Tummala, R. (2002). RF-system-on-package (SOP) for wireless communications. *IEEE Microwave Magazine*, *3*(1), 88–99. doi:10.1109/MMW.2002.990700

Liu, Y., Meng, H., Li, G., & Wang, X. (2008, August). Range-Velocity Estimation of Multiple Targets in Randomised Stepped-Frequency Radar. *Electronics Letters*, *44*(17), 1032–1034. doi:10.1049/el:20081608

Liu, Z. G., & Guo, Y. X. (2005). Compact low-profile dual band metamaterial antenna for body centric communications. *IEEE Antennas and Wireless Propagation Letters*, *14*, 863–866. doi:10.1109/LAWP.2014.2382586

Manna, S., Bose, T., & Bera, R. (2018). Wearable Antennas for Medical Application: A Review. In A. Kalam, S. Das, & K. Sharma (Eds.), *Advances in Electronics, Communication and Computing. Lecture Notes in Electrical Engineering* (Vol. 443). Springer. doi:10.1007/978-981-10-4765-7_13

Rachana, K., Singh, V. K., Singh, N. K., & Anurag, S. (2018). A Compact Pentagonal Textile Microstrip Antenna for Wide Band Application. Lecture Notes in Electrical Engineering, 443, 793-799.

Sharma, A., Chen, C. R., & Vu Lan, N. (2009). Solar-Energy Drying Systems: A Review. *Renewable & Sustainable Energy Reviews*, *13*(6-7), 1185–1210. doi:10.1016/j.rser.2008.08.015

Singh Nikhil, K., & Singh Vinod, K. (2018). A Compact Slotted Textile Patch Antenna for Ultra-wide Band Application. Lecture Notes in Electrical Engineering, 443, 53-59.

Tomar, S. S., Agrawal, N., & Bansal, P. (2017). Various Modulation techniques in symmetrical multilevel inverters. *International Conference on Energy, Communication, Data Analytics and soft Computing (ICECDS 2017),* 1322-1327.

Van Torre, P., Vallozzi, L., Rogier, H., & Verhaevert, J. (2010). Diversity textile antenna systems for firefighters. *Antennas and Propagation (EuCAP), Proceedings of the Fourth European Conference.*

Yadav & Singh. (2019). Design of U-shape with DGS circularly polarized wearable antenna on fabric substrate for WLAN and C-Band applications. *Journal of Computational Electronics, 18*(3), 1103–1109.

Iannacci, J., Huhn, M., Tschoban, C., & Pötter, H. (2016a). RF-MEMS technology for 5G: Series and shunt attenuator modules demonstrated up to 110 GHz. *IEEE Electron Device Letters*, *37*(10), 1336–1339. doi:10.1109/LED.2016.2604426

Iannacci, J., Huhn, M., Tschoban, C., & Pötter, H. (2016b). RF-MEMS technology for future mobile and high-frequency applications: Reconfigurable 8-bit power attenuator tested up to 110 GHz. *IEEE Electron Device Letters*, *37*(12), 1646–1649. doi:10.1109/LED.2016.2623328

Shekhar, S., Vinoy, K. J., & Ananthasuresh, G. K. (2018). Low-voltage high-reliability MEMS switch for millimeter wave 5G applications. *Journal of Micromechanics and Microengineering*, *28*(7), 075012. doi:10.1088/1361-6439/aaba3e

Cetiner, B. A., Qian, J. Y., Chang, H. P., Bachman, M., Li, G. P., & De Flaviis, F. (2003). Monolithic integration of RF MEMS switches with a diversity antenna on PCB substrate. *IEEE Transactions on Microwave Theory and Techniques*, *51*(1), 332–335. doi:10.1109/TMTT.2002.806521

Hong, W., Jiang, Z. H., Yu, C., Zhou, J., Chen, P., Yu, Z., ... Cheng, Y. (2017). Multibeam antenna technologies for 5G wireless communications. *IEEE Transactions on Antennas and Propagation*, *65*(12), 6231–6249. doi:10.1109/TAP.2017.2712819

Real, S. M., Turjo, R. R., Sobhan, F. F., & Ashraf, N. (2018). *Phased array antennas for mmWave frequency spectrum for 5G communication* (Doctoral dissertation). BRAC University.

McLevige, W. V., & Sokolov, V. (1980). Microwave switching with parallel-resonated GaAs FETs. *IEEE Electron Device Letters*, *1*(8), 156–158. doi:10.1109/EDL.1980.25271

Gaspari, R. A., & Yee, H. H. (1978, June). Microwave GaAs FET switching. In *1978 IEEE-MTT-S International Microwave Symposium Digest* (pp. 58-60). IEEE.

Schiffman, B. M. (1958). A new class of broad-band microwave 90-degree phase shifters. *I.R.E. Transactions on Microwave Theory and Techniques*, *6*(2), 232–237. doi:10.1109/TMTT.1958.1124543

Aghwariya, M. K., Agarwal, T., & Sharma, R. (2020). Frequency Scanning SIW Leaky Wave Horn Antenna for Wireless Application. In *International Conference on Intelligent Computing and Smart Communication 2019. Algorithms for Intelligent Systems*. Springer. 10.1007/978-981-15-0633-8_79

Aware, R., & Thorat, B. N. (2012). Solar Drying of Fruits and Vegetables. In Solar drying: Fundamentals, Applications and Innovations. Academic Press.

Gao, P., Hu, B., Tian, X.-L., Zhao, Q.-L., & Zhang, B.-T. (2017, April). Experimental study of a wearable aperture-coupled patch antenna for wireless body area network. *Microwave and Optical Technology Letters*, *59*(4), 761–766. doi:10.1002/mop.30408

Hall, P. S., & Hao, Y. (2006). *Antennas and Propagation for Body-Centric Wireless communications*. Artech House.

McMurray, W. (1971). The thyristor electronic transformer: A power converter using a high-frequency link. *IEEE Transactions on Industry and General Applications*, *IGA-7*(4), 451–457. doi:10.1109/TIGA.1971.4181326

Morelande, Mei, & Evans. (2013). *A Method of Target Detection*. cA Patent.

Najafi & Yatim. (2012). Design and Implementation of a new multilevel inverter topology. *IEEE Transactions on Industrial Electronics, 59*(11), 4148-4154.

Naresh, B., & Singh, V. K. (2020). Integration of RF Rectenna with thin film solar cell to power wearable electronics. *International Journal of Microwave and Wireless Technologies*. doi:10.100740009-020-00877-4

Saxena, Raizada, Gautam, & Khare. (2019). Efficient Rectenna Circuit for Wireless Power Transmission. In *Design and Optimization of Sensors and Antennas for Wearable Devices: Emerging Research and Opportunities*. IGI Global.

Singh, V. K., Dhupkariya, S., & Bangari, N. (2017). Wearable Ultra Wide Dual Band Flexible Textile Antenna for WiMax/WLAN Application. *International Journal of Wireless Personal Communications, 95*(2), 1075–1086.

Varadan, V. K., Vinoy, K. J., Jose, K. A., & Mems, R. (2003). *Their Applications*. Pennsylvania State University.

Garver, R. V. (1972). Broad-band diode phase shifters. *IEEE Transactions on Microwave Theory and Techniques*, *20*(5), 314–323. doi:10.1109/TMTT.1972.1127751

Glance, B. (1979). A fast low-loss microstrip PIN phase shifter. *IEEE Transactions on Microwave Theory and Techniques*, *27*(1), 14–16. doi:10.1109/TMTT.1979.1129551

White, J. F. (1974). Diode phase shifters for array antennas. *IEEE Transactions on Microwave Theory and Techniques*, *22*(6), 658–674. doi:10.1109/TMTT.1974.1128309

Kim, C. S., Lee, C. W., & Borer, J. R. (1966). Varactor S-band direct phase modulator. *IEEE Journal of Solid-State Circuits*, *1*(1), 45–51. doi:10.1109/JSSC.1966.1049755

Miao, M., Wu, G., Li, Z., Jin, Y., & Hao, Y. (2006, January). A bulk micromachined distributed digital microwave phase shifter with butterfly multilayer bridges and MAM capacitors. In *2006 1st IEEE International Conference on Nano/Micro Engineered and Molecular Systems* (pp. 479-483). IEEE. 10.1109/NEMS.2006.334804

Guclu, C., Cetintepe, C., Civi, O. A., Demir, S., & Akin, T. (2010, August). 35 GHz phased array antenna using DMTL phase shifters. In *2010 10th Mediterranean Microwave Symposium* (pp. 114-117). IEEE.

Hayden, J. S., & Rebeiz, G. M. (2002, June). A low-loss Ka-band distributed MEMS 2-bit phase shifter using metal-air-metal capacitors. In *2002 IEEE MTT-S International Microwave Symposium Digest (Cat. No. 02CH37278)* (Vol. 1, pp. 337-340). IEEE. 10.1109/MWSYM.2002.1011625

Barker, S., & Rebeiz, G. M. (1998). Distributed MEMS true-time delay phase shifters and wide-band switches. *IEEE Transactions on Microwave Theory and Techniques*, *46*(11), 1881–1890. doi:10.1109/22.734503

Rebeiz, G. M., Tan, G. L., & Hayden, J. S. (2002). RF MEMS phase shifters: Design and applications. *IEEE Microwave Magazine*, *3*(2), 72–81. doi:10.1109/MMW.2002.1004054

Kim, H. T., Park, J. H., Lee, S., Kim, S., Kim, J. M., Kim, Y. K., & Kwon, Y. (2002). V-band 2-b and 4-b low-loss and low-voltage distributed MEMS digital phase shifter using metal-air-metal capacitors. *IEEE Transactions on Microwave Theory and Techniques*, *50*(12), 2918–2923. doi:10.1109/TMTT.2002.805285

Agrawal, N., Bansal, P., & Umale, N. (2020). THD Analysis of New Multilevel Inverter Topology with Different Modulation Techniques. In *International Conference on Intelligent Computing and Smart Communication 2019. Algorithms for Intelligent Systems*. Springer. 10.1007/978-981-15-0633-8_117

Al-Hourani, Evans, Farrell, Moran, Martorella, Kandeepan, Skafidas, & Parampalli. (2016). Millimeter-wave CMOS Radar Systems and Techniques. *Academic Press Library in Signal Processing, 7.*

Bowers, J. C., Garrett, S. J., Nienhaus, H. A., & Brooks, J. L. (1980, June). *A solid state transformer. In 1980 IEEE Power Electronics Specialists Conference*. IEEE.

Katehi, L. P. (1995). Si micromachining in high-frequency applications. *CRC Handbook on Si Micromachining*, 1771-1787.

Liu, C., Guo, Y. X., & Xiao, S. (2012). Compact dual-band antenna for implantable devices. *IEEE Antennas and Wireless Propagation Letters, 11*, 1508–1511. doi:10.1109/LAWP.2012.2233705

Paracha, K. N., Abdul Rahim, S. K., Soh, P. J., & Khalily, M. (2019). Wearable Antennas: A Review of Materials, Structures, and Innovative Features for Autonomous Communication and Sensing. *IEEE Access: Practical Innovations, Open Solutions*, *7*, 56694–56712. doi:10.1109/ACCESS.2019.2909146

Saxena, A., & Singh, V. K. (2018a). A Moon-Strip Line Antenna for Multi-Band Applications at 5.44 GHz Resonant Frequency. *4th International Conference on Advances in Electrical, Electronics, Information, Communication and Bio-Informatics (AEEICB-18).*

Singh, N., Singh, A. K., & Singh, V. K. (2015a). Design & Performance of Wearable Ultra Wide Band Textile Antenna for Medical Applications. *Microwave and Optical Technology Letters, 57*(7), 1553-1557.

Thakre, V. V., & Singhal, P. K. (2009). Band width analysis by introducing slots in microstrip antenna design using ANN. *Progress in Electromagnetics Research M. Pier M*, *9*, 107–122. doi:10.2528/PIERM09093002

Visvale, G. L. (2012). Classification and Selection of Solar Dryers. In Solar drying: Fundamentals, Applications and Innovations. Academic Press.

Xu & Li. (2012). A Dual band Microstrip Antenna for Wearable Application. *IEEE (ISAPE) Conference Proceeding*, 109-112.

Zhao, X. G., Yang, J. H., Fu, Y., Yang, D., Xu, Q., Yu, L., & Zhang, L. (2017). Design of lead-free inorganic halide perovskites for solar cells via cation-transmutation. *Journal of the American Chemical Society*, *139*(7), 2630–2638. doi:10.1021/jacs.6b09645 PMID:28112933

Dey, S., Koul, S. K., Poddar, A. K., & Rohde, U. L. (2017). Ku to V-band 4-bit MEMS phase shifter bank using high isolation SP4T switches and DMTL structures. *Journal of Micromechanics and Microengineering*, *27*(10), 105010. doi:10.1088/1361-6439/aa8071

Dey, S., Koul, S. K., Poddar, A. K., & Rohde, U. L. (2018). Reliable and compact 3-and 4-bit phase shifters using MEMS SP4T and SP8T switches. *Journal of Microelectromechanical Systems*, *27*(1), 113–124. doi:10.1109/JMEMS.2017.2782780

Topalli, K., Civi, Ö. A., Demir, S., Koc, S., & Akin, T. (2008). A monolithic phased array using 3-bit distributed RF MEMS phase shifters. *IEEE Transactions on Microwave Theory and Techniques*, *56*(2), 270–277. doi:10.1109/TMTT.2007.914377

Pomona, I., Farinelli, P., Microtech, R. F., Margesin, B., & Meneghesso, G. (2011). RF-MEMS Phase Shifters for Phased Array Antennas. In Nanotech Italy 2011 (No. 13, pp. 37-39). ITA.

Bhosale, P. S., & Kolhare, N. R. (2013). A review paper on design of distributed RF MEMS phase shifter for aerospace communication. *International Journal of Engineering Trends and Technology*, *4*(3), 284–288.

Qian, J. Y., Li, G. P., & De Flavis, F. (2000, June). A parametric model of MEMS capacitive switch operating at microwave frequencies. In *2000 IEEE MTT-S International Microwave Symposium Digest (Cat. No. 00CH37017)* (Vol. 2, pp. 1229-1232). IEEE. 10.1109/MWSYM.2000.863581

Kanthamani, S., Mohan, S., Raju, S., Abhaikumar, V., & Mohan, V. (2009). Meshless Analysis of Radio Frequency Microelectromechanical Systems Shunt Switch. *Defence Science Journal*, *59*(6), 622–626. doi:10.14429/dsj.59.1567

Lacroix, B., Pothier, A., Crunteanu, A., & Blondy, P. (2008, October). Phase shifter design based on fast RF MEMS switched capacitors. In *2008 European Microwave Integrated Circuit Conference* (pp. 478-481). IEEE. 10.1109/EMICC.2008.4772333

Varian, K., & Walton, D. (2002). A 2-bit RF MEMS phase shifter in a thick-film BGA ceramic package. *IEEE Microwave and Wireless Components Letters*, *12*(9), 321–323. doi:10.1109/LMWC.2002.803201

Sengupta, S., & Kiran, K. U. (2015). Miniatured phase shifter. *Indian Journal of Science and Technology*, *8*(19), 1–7. doi:10.17485/ijst/2015/v8i19/76867

Brown, E. R. (1999, December). Architectural issues in RF-MEMS based reconfigurable apertures. *NSF RF Micromachining MEMS Technol. Wireless Commun. Syst. Workshop*.

Cortecchia, D., Dewi, H. A., Yin, J., Bruno, A., Chen, S., Baikie, T., Boix, P. P., Grätzel, M., Mhaisalkar, S., Soci, C., & Mathews, N. (2016). Lead-free MA2CuCl x Br4–x hybrid perovskites. *Inorganic Chemistry*, *55*(3), 1044–1052. doi:10.1021/acs.inorgchem.5b01896 PMID:26756860

Goldberg, A. F., Kassakian, J. G., & Schlecht, M. F. (1987, June). *Issues related to 1–10 MHz transformer design. In 1987 IEEE Power Electronics Specialists Conference*. IEEE.

Jeng, S.-L., Chieng, W.-H., & Lu, H.-P. (2014, April). Estimating Speed Using a Side-Looking Single-Radar Vehicle detector. *IEEE Transactions on Intelligent Transportation Systems*, *15*(2), 607–614. doi:10.1109/TITS.2013.2283528

Lee, S., Kim, U., Kwon, K., Seo, W., & Choi, J. (2012). Design of on-body antenna for wireless body area network. In *Advance Communication Technology* (pp. 300–303). ICACT.

Ooi, H.S. (2006). Solar Assisted Crop Dryers for Paddy, Flowers and Kenaf. In *Solar Assisted Drying Systems, Innovative Technologies for Agricultural and Marine Products*. Islamic Education, Scientific and Cultural Organization.

Poornima, S., Singh, V. K., Archana, L., & Bhoi, A. K. (2018). Design and Analysis of Microstrip Antenna Using Multilayer Feed-Forward Back-Propagation Neural Network (MLPFFBP-ANN). *Lecture Notes in Electrical Engineering, 462*, 393–398. doi:10.1007/978-981-10-7901-6_43

Saxena, Singh, Mohini, Bhardwaj, Chae, Sharma, & Bhoi. (2018). Rectenna Circuit at 6.13 GHz to operate the sensors devices. *International Journal of Engineering and Technology, 7*(2.33), 644-646.

Singh, N., Singh, A. K., & Singh, V. K. (2015b). Ultra Wide Band Textile Antenna for Medical Applications. *Open Engineering, 5*(1), 117-119.

Wu, C.-M., Syu, J.-W., & Liu, W.-C. (2015). Dual-band slotted patch antenna with defective ground forWLAN/WiMAX applications. *Progress in Electromagnetics Research Letters, 53*, 1–6. doi:10.2528/PIERL15022503

Jahanbakht, M., Naser-Moghadasi, M., & Lotfi-Neyestanak, A. A. (2008). Fractal beam KU-band MEMS phase shifter. *Progress In Electromagnetics Research, 5*, 73–85. doi:10.2528/PIERL08101703

Du, Y., Bao, J., Wu, W., Tu, C., & Mu, P. (2010, July). Design of a novel 5-bit distributed MEMS phase shifter. In *2010 International Conference on Communications, Circuits and Systems (ICCCAS)* (pp. 706-709). IEEE.

Chakraborty, A., Kundu, A., Dhar, S., Maity, S., Chatterjee, S., & Gupta, B. (2011, September). Compact K-band distributed RF MEMS phase shifter based on high-speed switched capacitors. In *2011 11th Mediterranean Microwave Symposium (MMS)* (pp. 25-28). IEEE. 10.1109/MMS.2011.6068521

Chakraborty, A., Kundu, A., Chatterjee, S., & Gupta, B. (2012). *Design of Compact RF MEMS Phase Shifters Based on Space-filling Curves*. Academic Press.

Chakraborty, A., & Kar, A. K. (2016). Utilization of Fractal Geometry for Phase Shifter Implementation. In *Proceedings of the International Conference on Signal, Networks, Computing, and Systems* (pp. 317-323). Springer. 10.1007/978-81-322-3589-7_35

Chakraborty, A., & Gupta, B. (2016a). Development of compact 180° phase shifters based on MEMS technology. *Sensors and Actuators. A, Physical, 247*, 187–198. doi:10.1016/j.sna.2016.05.046

Chakraborty, A., & Gupta, B. (2016b). Paradigm phase shift: RF MEMS phase shifters: An overview. *IEEE Microwave Magazine, 18*(1), 22–41. doi:10.1109/MMM.2016.2616155

Chakraborty, A., & Gupta, B. (2017). Utility of RF MEMS miniature switched capacitors in phase shifting applications. *AEÜ. International Journal of Electronics and Communications*, *75*, 98–107. doi:10.1016/j.aeue.2017.03.011

Niedermeier, R., Reinhardt, K., & Sanders, P. (1997, September). Towards optimal locality in mesh-indexings. In *International Symposium on Fundamentals of Computation Theory* (pp. 364-375). Springer. 10.1007/BFb0036198

Haverkort, H., & van Walderveen, F. (2010). Locality and bounding-box quality of two-dimensional space-filling curves. *Computational Geometry*, *43*(2), 131–147. doi:10.1016/j.comgeo.2009.06.002

Gradzki, P. M., Jovanovic, M. M., & Lee, F. C. (1990, March). Computer-aided design for high-frequency power transformers. In *Fifth Annual Proceedings on Applied Power Electronics Conference and Exposition* (pp. 336-343). IEEE. 10.1109/APEC.1990.66427

Grilo, M., & Correra, F. S. (2015, July). Rectangular patch antenna on textile substrate fed by proximity coupling. *Journal of Microwaves, Optoelectronics and Electromagnetic Applications*, *14*, 103–112.

Hii, C. L., & Law, C. L. (2012). Solar Drying of Major Commodity Products. In Solar drying: Fundamentals, Applications and Innovations. Academic Press.

Ke, W., & Kanatzidis, M. G. (2019). Prospects for low-toxicity lead-free perovskite solar cells. *Nature Communications*, *10*(1), 965. doi:10.103841467-019-08918-3 PMID:30814499

Lin, & Li. (2016). Design of an FMCW radar baseband signal processing system for automotive application. *SpringerPlus*.

Liu, Z. G., & Guo, Y. X. (2013). Dual band low profile antenna for body centric communications. *IEEE Transactions on Antennas and Propagation*, *61*(4), 2282–2285. doi:10.1109/TAP.2012.2234071

Lui, K. W., Murphy, O. H., & Toumazou, C. (2013, July). wearable wideband circularly polarized textile antenna for effective power transmission on a wirelessly-powered sensor platform. *IEEE Transactions on Antennas and Propagation*, *61*(7), 3873–3876. doi:10.1109/TAP.2013.2255094

Petersen, K. E. (1979). Micromechanical membrane switches on silicon. *IBM Journal of Research and Development*, *23*(4), 376–385. doi:10.1147/rd.234.0376

Sharma, Yadav, & Singh. (2018). Design of Circularly Polarized Antenna with Different Iterations for UWB Applications. Lecture Notes in Electrical Engineering, 443, 441-447.

Singh Nikhil, Singh Vinod, Saxena, Bhoi, Garg, & Sherpa. (2018). A Compact Slotted Textile Patch Antenna for Ultra-wide Band Application. Lecture Notes in Electrical Engineering, 443, 53-59.

Singh, V. K., Ali, Z., Ayub, S., & Singh, A. K. (2014a). Bandwidth Optimization of Compact Microstrip Antenna for PCS/DCS/Bluetooth Application. *Open Engineering, 4*(3), 281-286.

Singhal, P. K. (2013). *Dual Director Microstrip Planar Yagi-Uda Antenna for X-Band* (Vol. 3). IJECCT.

Bathe, K. J., Ramm, E., & Wilson, E. L. (1975). Finite element formulations for large deformation dynamic analysis. *International Journal for Numerical Methods in Engineering*, *9*(2), 353–386. doi:10.1002/nme.1620090207

Nabors, K., & White, J. (1991). FastCap: A multipole accelerated 3-D capacitance extraction program. *IEEE Transactions on Computer-Aided Design of Integrated Circuits and Systems*, *10*(11), 1447–1459. doi:10.1109/43.97624

Phillips, J. R., & White, J. (1994, November). A precorrected-FFT method for capacitance extraction of complicated 3-D structures. In ICCAD (Vol. 94, pp. 268-271). Academic Press.

Yie, H., Cai, X., & White, J. (1994, June). Convergence properties of relaxation versus the surface-Newton generalized-conjugate residual algorithm for self-consistent electromechanical analysis of 3-D micro-electro-mechanical structures. In *Proceedings of International Workshop on Numerical Modeling of processes and Devices for Integrated Circuits: NUPAD V* (pp. 137-140). IEEE. 10.1109/NUPAD.1994.343472

Gilbert, J. R., Legtenberg, R., & Senturia, S. D. (1995, February). 3D coupled electro-mechanics for MEMS: applications of CoSolve-EM. In *Proceedings IEEE Micro Electro Mechanical Systems* (p. 122). IEEE. 10.1109/MEMSYS.1995.472542

Bachtold, M., Korvink, J. G., Funk, J., & Baltes, H. (1995, December). New convergence scheme for self-consistent electromechanical analysis of iMEMS. In *Proceedings of International Electron Devices Meeting* (pp. 605-608). IEEE. 10.1109/IEDM.1995.499294

Aluru, N. R., & White, J. (1996, June). Direct-Newton finite-element/boundary-element technique for micro-electro-mechanical analysis. In *Tech. Digest, Solid-State Sensor and Actuator Workshop* (pp. 54-57). Academic Press.

Aluru, N. R., & White, J. (1997). An efficient numerical technique for electrochemical simulation of complicated microelectromechanical structures. *Sensors and Actuators. A, Physical*, *58*(1), 1–11. doi:10.1016/S0924-4247(97)80218-X

Tentzeris, E. M. (2002, October). Adaptive modeling and design of highly integrated 3D microwave-millimeter wave radio front-ends. In 24th Annual Technical Digest Gallium Arsenide Integrated Circuit (GaAs IC) Symposium (pp. 109-112). IEEE.

Fei, T. (2017). A Novel Target Separation Algorithm Applied to The Two- Dimensional Spectrum for FMCW Automotive Radar Systems. *IEEE International Conference on Microwaves, Antennas, Communications and Electronic Systems*. 10.1109/COMCAS.2017.8244791

Hao, F., Stoumpos, C. C., Cao, D. H., Chang, R. P., & Kanatzidis, M. G. (2014). Lead-free solid-state organic–inorganic halide perovskite solar cells. *Nature Photonics*, *8*(6), 489–494. doi:10.1038/nphoton.2014.82

Jain, A. K., Kanaujia, B. K., Dwari, S., Pandey, G. P., & Singh, D. K. (2019, June). Wideband circularly polarized magnetoelectric dipole antenna with I-slot for C-band applications. *Journal of Computational Electronics*, *18*(2), 660–670. doi:10.100710825-018-01295-y

Katehi, L. P., Harvey, J. F., & Brown, E. (2002). MEMS and Si micromachined circuits for high-frequency applications. *IEEE Transactions on Microwave Theory and Techniques*, *50*(3), 858–866. doi:10.1109/22.989969

Naresh, B., Singh, V. K., & Bhargavi, V. (2017). Dual band RF energy harvester for wearable electronic technology. *2017 Third International Conference on Advances in Electrical, Electronics, Information, Communication and Bio-Informatics (AEEICB)*, 274-277. 10.1109/AEEICB.2017.7972428

Rabinowitz, M. (2003). *Improved Power System of the Future.* arXiv preprint physics/0304083.

Rahul, Singh, Singh, Prakash, & Bhoi. (2018). Multi Resonant Textile Antenna with Partial Ground for Multiband Applications. Lecture Notes in Electrical Engineering, 462, 359-367.

Saxena, A., & Singh, V. K. (2018b). A Watch E-Cut Shape Textile Antenna for WB Applications. *Journal of Microwave & Technologies*, *5*(1), 29–32.

Singh, Ali, Singh, & Ayub. (2013). Dual Band Triangular Slotted Stacked Microstrip Antenna for Wireless Applications. *Open Engineering, 3*(2), 221-225.

Zhu, X.-Q., Guo, Y.-X., & Wu, W. (2016). A novel dual-band antenna for wireless communication applications. *IEEE Antennas and Wireless Propagation Letters*, *15*, 516–519. doi:10.1109/LAWP.2015.2456039

Anderson, Ghafurian, & Gharavi. (2018). *Smart Grid: The Future of the Electric Energy System.* Research Gate.

Asif, M., Muhammad, A., & Irfan, A. M. (2008). Design and implementation of AMR Smart Grid System. *IEEE Canada Electric Power Conference.*

Balanis, C. A. (2004). *Antenna Theory: Analysis and Design.* John Wiley and Sons.

Brown, H. E., & Suryanarayanan, S. (2009). A survey seeking a definition of a smart distribution system. *North American power symposium'09*, 1–7. 10.1109/NAPS.2009.5484078

Dhupkariya, S., & Singh, V. K. (2015). Textile Antenna for C-Band Satellite Communication Application. *Journal of Telecommunication, Switching Systems and Networks, 2*(2), 20-25.

Dubey, R., Singh, V. K., Bhoi, A. K., & Ali, Z. (2018). Realization of trophy shaped flexible wearable antenna based on foam substrate. *International Journal of Engineering & Technology, 7*, 222-224.

Gharavi, H., & Ghafurian, R. (2011). Smart Grid: The Electric Energy System of the Future. *Proceedings of the IEEE, 9.*

Giordano, V., & Bossart, S. (2012). *Accessing smart grid benefits and impacts: EU and U.S initiatives.* Academic Press.

Hassan, R., & Radman, G. (2010). Survey on smart grid. *IEEE Southeast on 2010*, 210–213.

Indrajeet. (2014). Smart Grid Technology: Application and Control. *International Journal of Advanced Research in Electrical, Electronics and Instrumentation Engineering, 3*(5).

Ipakchi, A., & Albuyeh, F. (2009). Grid of the future. *IEEE Power & Energy Magazine, 7*(2), 52–62. doi:10.1109/MPE.2008.931384

Katz, J. (2008). Educating the Smart Grid. *IEEE Energy 2030.*

Khare, B. B., Singh, V. K., & Saxena, A. (2020). Sigma-Structured Microstrip Antenna for Harvesting Energy for Low-Power Devices. In *International Conference on Intelligent Computing and Smart Communication 2019. Algorithms for Intelligent Systems.* Springer. 10.1007/978-981-15-0633-8_73

Kushwaha, R., Singh, V. K., Singh, N. K., Saxena, A., & Sharma, D. (2018). A Compact Pentagonal Textile Microstrip Antenna for Wide Band Application. Advances in Electronics, Communication and Computing. *Lecture Notes in Electrical Engineering, 443*, 793–799.

Loni, Ayub, & Singh. (2014). Performance analysis of Microstrip Patch Antenna by varying slot size for UMTS application. *IEEE Publication, Communication Systems and Network Technologies (CSNT-2014).*

Loni, J., & Singh, V. K. (2015). Development of Bandwidth Enhanced Microstrip Patch Antenna for UMTS Application. *Journal of Microwave Engineering & Technologies, 2*(1), 1-7.

Ma, X. (2020). *Smart Grid.* https://www.cse.wustl.edu/~jain/cse574-10/ftp/grid/

Pipattanasomporn, A., & Rahman, A. (2009). Multi-Agent Systems in a Distributed Smart Grid: Design and Implementation. *Proc. IEEE PES 2009 Power Systems Conference and Exposition.* 10.1109/PSCE.2009.4840087

Rahimi, F., & Ipakchi, A. (2010). Demand response as a market resource under the smart grid paradigm. *IEEE Transactions on Smart Grid, 1*(1), 82–88. doi:10.1109/TSG.2010.2045906

Rawat, S., & Sharma, K. K. (2014). A compact broadband microstrip patch antenna with defected ground structure for C-band applications. *Central European Journal of Engineering, 4*(3), 287–292.

Samad, T., & Annaswamy, A. M. (2011). *The Impact of control technology- Control for renewable energy and Smart Grid.* www.ieeecss.org

Singh & Naresh. (2015). Multi Resonant Microstrip Antenna with Partial Ground for Radar Application. *Journal of Telecommunication, Switching Systems and Networks, 2*(1), 1-5.

Singh & Singh. (2015). Compact Circular Slotted Microstrip Antenna for Wireless Communication Systems. *Journal of Microwave Engineering & Technologies, 1*(1), 7-14.

Singh, A. K., & Kabeer, R. A. (2013). Performance Analysis of First Iteration Koch Curve Fractal Log Periodic Antenna of Varying Angles. *Open Engineering, 3*(1), 51-57.

Singh, Ali, Ayub, & Singh. (2014). Bandwidth Optimization of Compact Microstrip Antenna for PCS/DCS/Bluetooth Application. *Open Engineering, 4*(3), 281-286.

Singh, M., & Singh, V. K. (2015). Rectangular Slot Loaded Circular Patch Antenna for WLAN Application. *Journal of Telecommunication, Switching Systems and Networks, 2*(1), 7-10.

Singh, N., Singh, A. K., & Singh, V. K. (2015). Design & Performance of Wearable Ultra Wide Band Textile Antenna for Medical Applications. *Microwave and Optical Technology Letters, 57*(7), 1553-1557.

Singh, V. K., Ali, Z., Ayub, S., & Singh, A. K. (2014). A wide band Compact Microstrip Antenna for GPS/DCS/PCS/WLAN Applications. In Intelligent Computing, Networking, and Informatics. Academic Press.

Singh, V. K., Dhupkariya, S., & Bangari, N. (2017). Wearable Ultra Wide Dual Band Flexible Textile Antenna for WiMax/WLAN Application. *International Journal of Wireless Personal Communications, 95*(2), 1075–1086.

Singh, N. K., & Singh, V. K. (2016). Textile Antenna for Microwave Wireless Power Transmission. *Procedia Computer Science.*

Souran, Safa, Moghadam, Ghasempour, & Heravi. (2015). *Smart Grid Technology in Power Systems*. Research Gate.

Srivastava, Ayub, & Singh. (2014). Dual Band Rectangular and Circular Slot Loaded Microstrip Antenna for WLAN/GPS/WiMax Applications. *IEEE Publication, Communication Systems and Network Technologies (CSNT-2014).*

Srivastava, R., Singh, V. K., & Ayub, S. (2015). Comparative Analysis and Bandwidth Enhancement with Direct Coupled C Slotted Microstrip Antenna for Dual Wide Band Applications. In Frontiers of Intelligent Computing: Theory and Applications. Academic Press.

Srivastava, S., Singh, V. K., Ali, Z., & Singh, A. K. (2013). Duo Triangle Shaped Microstrip Patch Antenna Analysis for WiMAX lower band Application. *Procedia Technology, 10*, 554-563.

Uslar, M., Rohjansand, S., Bleiker, R., Gonz'alez, J., Specht, M., Suding, T., & Weidelt, T. (2010). Survey of smart grid standardization studies and recommendations—Part 2. *IEEE PES, 10*, 1–6. doi:10.1109/ISGTEUROPE.2010.5638886

Xu, L., & Li, J. L.-W. (2012). A Dual band Microstrip Antenna for Wearable Application. *IEEE (ISAPE) Conference Proceeding*, 109-112.

Yadav & Singh. (n.d.). Design of U-shape with DGS circularly polarized wearable antenna on fabric substrate for WLAN and C-Band applications. *Journal of Computational Electronics.*

Yadav, A., Kumar Singh, V., Kumar Bhoi, A., Marques, G., Garcia-Zapirain, B., & de la Torre Díez, I. (2020). Wireless Body Area Networks: UWB Wearable Textile Antenna for Telemedicine and Mobile Health Systems. *Micromachines, 11*(6), E558.

Yadav, A., Singh, V.K., Yadav, P., Beliya, A.K., Bhoi, A.K., & Barsocchi, P. (2020). Design of Circularly Polarized Triple-Band Wearable Textile Antenna with Safe Low SAR for Human Health. *Electronics, 9*, 1366.

Yu, Y., & Luan, W. (2009). Smart grid and its implementations. *CSEE, 29*(34), 1–8.

Zahran, M. (2013). *Smart grid technology, vision management and control. WSEAS Transactions on Systems*, 12.

# About the Contributors

**Vinod Kumar Singh** is working as Professor and Head in Electrical Engineering Department at S.R. Group of Institutions, Jhansi UP. He received PhD in Electronics and Communication Engineering from BU Rajasthan in 2013 and M. Tech. in Digital Communication System from B.I.E.T., Jhansi in 2009.He completed his B.Tech in Electrical Engineering from IET MJP Rohilkhand University, Bareilly in 2001. He is having about Eighteen years of experience in the field of Electrical/ Electronics Engineering. India. He is a senior member of International Association of Computer Science and Information Technology (IACSIT) and International Association for the Engineers & Computer Scientists (IAENG).Prof. Singh has published more than 200 research papers in the renowned International Journals such as IEEE, Springer and Willey.

* * *

**Nikhil Agrawal** received B.Tech degree with honors in Electrical Engineering in 2014 and Master of Engineering in Industrial systems & Drives From M.I.T.S. Gwalior in 2017. He awarded with M.I.T.S. Gold Medal 2017 for securing first position in M.E. After M.E., he joined the Ujjain Engineering College, Ujjain as a lecturer in Electrical Engineering Department, After that he joined S.R. group of Institutions, Jhansi as an Assistant Professor in Electrical Engineering department. He has published research papers in IEEE, Springer Conferences, International journal and national conferences. He is reviewer of renowned journal, springer conferences such as ICSC 2019, ICAEEC 2019.

**Bharat Bhushan Khare** is pursuing Master of Engineering in Power System From U.I.T. RGPV Bhopal, MP and completed B.Tech. in Electrical Engineering from S R Group of Institutions Jhansi. He is currently working on designing of power electronics converters and wireless energy transfer.

**Akanksha Lohia** received B.Tech degree with honours in Electronics engineering in 2014 and M.E. with honours in communication control & Networking from MITS Gwalior in 2017. After M.E. she joined the SR Group of institution Jhansi as an Assistant Professor in Electronics Engineering Department. Ms Lohia has published research papers in renowned journals and conferences.

**Sanjeev Sharma** received his B.Tech degree from IET, Barielly, he received his M.Tech degree from RKNEC, RTMNU, India, he received his PG Diploma in VLSI from CDAC, and completed his Ph.D from RTMNU, he has over 7 years of industrial experience, working at CDAC and 12 years of teaching experience in reputed engineering college. At present he is a professor and HOD at NHCE, Bangalore. His overall 40 Papers published in international and national conference on selected areas in communication and embedded systems, his research interests include artificial intelligence, wireless communication, embedded systems VLSI, antennas, and soft-core processor.

**Manimegalai T.** received the Bachelor of Engineering degree in Electronics and Communication Engineering from Bharathidasan University, Tamilnadu, India, in 1996. She received the Master of Engineering degree in Applied Electronics from Madurai Kamaraj University, India in 1998. She is working as Associate Professor in SRM Institute of Science and Technology, Chennai, India. She Completed her PhD in the Department of Electronics and Communication Engineering, SRMIST, Chennai, India. Her research interests are Ultra-Wideband, wireless sensor networks, image processing and MIMO-OFDM systems.

**Reji V.** has completed her B.E in Electronics and Communication Engineering from Manonmanium Sundharanar University, Tamil Nadu, India in 2003. She received the masters of engineering Degree in Communication Systems at SRM University, India, 2011. She is working as Assistant Professor at SRM Institute of Science and Technology, Chennai, India. Her research interest in Antennas and RF design.

# Index

# I



# J



# K



# L



# M



# N



# P



# R



# S



# T



# V



# W



# X

www.ingramcontent.com/pod-product-compliance
Lightning Source LLC
LaVergne TN
LVHW081519300926
849161LV00006B/1225